Selected Titles in This Series

D0875155

CBMS

Conference Board of the Mathematical Sciences

Issues in Mathematics Education

Volume 6

Research in Collegiate Mathematics Education. II

Jim Kaput
Alan H. Schoenfeld
Ed Dubinsky
Editors

Thomas Dick, *Managing Editor*

American Mathematical Society
Providence, Rhode Island
in cooperation with
Mathematical Association of America
Washington, D. C.

1991 *Mathematics Subject Classification.* Primary 00-XX, 92-XX.

ISBN 0-8218-0382-4
ISSN 1047-398X

CONTENTS

PREFACE

Welcome to the second volume in the *Research in Collegiate Mathematics Education (RCME)* series. For a general introduction to the series, see the preface to *RCME I*; and for a general introduction to the field see the first chapter by Alan Schoenfeld in that volume. In this preface we will simply introduce the papers, but with an orientation to future research. Since the field of research in collegiate mathematics education is very young, most studies will raise many more questions than they answer. Some will open wide new avenues for investigation, and a few will offer an avenue already well paved and traveled. Such variation is the case with the papers in this volume as well as those in preparation for the third volume. They exhibit large diversity in methods and purposes, ranging from historical studies, to theoretical examinations of the role of gender in mathematics education, to practical evaluations of particular practices and circumstances.

Parallel to *RCME I*, this collection begins with a paper that attends to methodology and closes with a list of questions. The lead-off paper, from a research group led by Ed Dubinsky, describes a distinctive approach to research on key concepts in the undergraduate mathematics curriculum. This approach is distinguished from others in several ways, key among these being its integration of research and instruction. The method begins with a careful "genetic" (in the sense of Piaget) preliminary analysis of the concept's epistemological structure. This analysis is applied to build a first approximation instructional treatment intended to enable a student to build this knowledge structure. A variety of data are gathered and analyzed regarding the cognitive impact of the intervention with a view to refining both the theoretical analysis and the instructional intervention in preparation for another instructional cycle. The typical instructional treatment involves interactive computer environments and some collaborative learning; and the research data typically involve a mix of open-response written tests and clinical interviews. Among the topics addressed through this approach are functions, group theory, calculus, discrete mathematics. This work has begun to reveal some underlying concept-development invariants in the form of a sequence of progressions that moves from action to process to object status for a number of mathematics concepts. A general question that can be asked of this framework concerns its scope of applicability—How much mathematics learning, and what kinds of learning, can be studied in this way? And what does not fit into this picture?

David Dennis and Jere Confrey look carefully at the way mathematical knowledge was built in the time of Wallis—how did mathematicians enlarge their conceptions of exponent from positive integers to continuous exponents? In so doing, they uncover a wealth of conceptual integrations of important ideas—powers, indices, exponents, area, ratios, limits of ratios, negative numbers and continuous functions—which in turn were explored across different representations—table arithmetic, geometry, and algebra. The interactions of these ideas across representations turns out to be an important factor in establishing the certainty of relationships and the reliability of constructions—if it makes sense in a second representation, then its epistemological status is greatly enhanced. Another point made by the authors is that the strongly empirical and heuristic methods by which ideas were built, examined and tested, methods that predominated up through the work of Euler and strongly exemplified by Wallis, are now obscured and disacknowledged in favor of the formal, algebraically-based constructions and proofs that followed. The paper can be read on at least two levels, one as careful history, and the other as a source of ideas for teaching mathematics and reflecting on our current (none-too-successful) approaches. As history, it raises the controversial question of the role of Alhazen in Wallis' work and counters the frequently offered assessment of Wallis as one who was reckless in jumping to conclusions, since the authors illustrate how he inevitably took his conjectures, usually arithmetic, to another representational realm for confirmation, usually geometric. As education, it raises serious questions about how much we take for granted and, in so doing, ignore, in our algebraically dominated approach to school mathematics. We are led to wonder how much richer and more powerful our students' knowledge might be if our curriculum and teaching included some of the ideas and approaches that historically lie behind the formalisms and methods based upon them. More generally, the question of the connections between history and education is raised anew.

The paper by Rina Zazkis and Ed Dubinsky on interpreting dihedral groups actually begins with conflicting interpretations of D_4 that turned up in student interviews, but eventually reaches much farther, to surprisingly widespread intrinsic contravariance relations between choices of notations for objects and notations for transformations of those objects. A geometric approach to representing D_4 as rigid actions on a square vs. an "analytical" approach based on permutations (i.e., as a subset of S_4), lead, in a natural way, to a contravariant relation between the operations of D_4 and those of S_4. This fact is seldom given explicit attention in Abstract Algebra, although many years ago, Mac Lane and Birkhoff identified the issue in terms of the "alias/alibi" distinction - does one name objects in terms of the ways their names change or the way their locations change under transformations? The authors offer a general model for interpreting the phenomenon based on how one interprets transformations of the "coordinate system" in which the objects are described. Their pedagogical recommendation is to be consistent in one's interpretation, and perhaps even go farther in being explicit about how to translate (via anti-isomorphisms) between interpretations. Armed with a detailed analysis of this example, they then revisit the alias/alibi distinction in two-space, in the context of quadratic forms, matrices of linear transformations, and connections between vector space automorphisms and matrices of linear transformations. They expose similarities as well as irreducible differences. In so doing, they raise to a new level of explicitness a range of phenomena with roots in basic abstract algebra. We wonder what other sorts of subtle phenomena crossing familiar territory might be uncovered by looking closely at student confusions?

The paper by Annette Leitze turns on a very practical issue—what factors, especially affective ones, enter into a student's decision to major in mathematics? She briefly reviews the literature on factors influencing student interest in the study of mathematics before describing her extensive empirical study of potential majors at a large public research university. Part of the study was survey-based, where she carefully compared major and non-major pairs matched for calculus professor, course grade, and gender. The remainder of the study used a small subsample who participated in a lengthy semi-structured interview. Overall, she found substantial differences in perceived usefulness of mathematics—much lower among majors, who had great difficulty even identifying professions that use the mathematics that they were to be learning. The data on student perception of the difficulty of the subject, its relations to their own sense of mathematical ability, and enjoying the subject (often conflated with appreciation of their instructors), are complex, and simple conclusions are not warranted. However, one aspect of majors' view of the subject that seemed consistent across both majors and non-majors was the view that mathematics is an "asocial subject," a subject that one studies or does alone. This study raises interesting questions about the common practices of mathematics instructors and departments and how they fit or fail to fit with students' expectations regarding the subject matter, how it may be learned, and how it might eventually be a part of their lives. The paper offers a different lens on the phenomena discussed by Marcia Linn and Cathy Kessel in the next paper.

Linn and Kessel provide a wide-ranging review of enrollment and persistence trends among mathematics majors: who stays, who leaves, and for what reasons? The issues are made clear in the opening sentences of the paper: "Two-thirds of the students who plan to study mathematics in college eventually choose other fields. Furthermore, gender but not grades or test scores predicts who switches out of mathematics." Linn and Kessel offer some data that may be surprising— e.g., that on average, the mathematics grades of "switchers" are every bit as good as those of students who remain mathematics majors, and that factors other than grades are largely responsible for students' moving out of mathematics. They review students' perceptions of their mathematics experiences, the predictive validity of test scores, and student performance data. The paper offers a set of implications for instruction, and research-grounded recommendations for making mathematics instruction more enfranchising for all students.

The paper by Sandra L. Burmeister, Patricia Ann Kenney, and Doris L. Nice explores the effectiveness of Supplemental Instruction (SI), a particular model of out-of-class learning assistance widely used in institutions of higher education, in a range of college courses dealing with algebra, calculus, and statistics. The authors describe the SI model and then analyze data gathered from 45 different institutions of higher education where the model was implemented for various "gatekeeper" courses in collegiate mathematics. Overall, the data indicate positive effects, with students who underwent SI instruction receiving higher grades than a comparable pool of students who did not. The paper concludes with the discussion of a series of issues related to the use of SI which require further investigation.

Kyungmee Park and Kenneth J. Travers offer a comparative study of a standard college first-year calculus course and a computer-based course, Calculus & Mathematica. Like many of the calculus-related papers in *RCME I*, this paper is of interest for at least two reasons. First, it elaborates the goals of a non-standard course and provides ways of documenting the effects of the instruction. As recent documents on assessment make clear (see, e.g., *Student Assessment in Calculus: NSF Working Group on Assessment in Calculus*, in press), the instructional goals of new courses differ in many ways from those of traditional courses. Hence, meaningful assessment of the new courses demands the gathering and analysis of data that correspond to these new instructional goals. Second, the paper introduces some research methods to readers (e.g., the use of concept maps and ways of quantifying them) that are of interest in their own right. This is yet another step in the development of a full complement of techniques for understanding and documenting student understanding.

The paper by Alvin Baranchik and Barry Cherkas looks at student strategies (deliberate or not) in taking multiple-choice placement exams. Evaluation of an individual's knowledge in mathematics, whether it be for the purpose of placement, grading, or giving awards, is a topic of major interest today. Although many people are questioning our traditional methods of testing, it seems clear that the multiple-choice exam with its apparent objectivity and cost-effectiveness for processing large numbers of individuals, will remain with us for some time to come. Given that, the issue raised by this paper can be rather disturbing. The authors consider the effect on performance on such an exam by factors that have little to do with what the student knows or understands, and may even have uneven impact depending on student characteristics such as ethnicity, gender, first language and age. We are reminded in this paper that when you take a multiple-choice exam it makes a difference whether you guess, omit questions, or simply not finish. The difference depends on how the exam is scored, that is, by merely counting the number correct, by subtracting the wrong answers from the right ones, or some other formula. What we learn in this report is that the choice of strategy by test takers varies with their ethnicity, gender, first language and age in ways that cannot be attributed to differences in mathematical ability. This is rather troubling when we think of the potential consequences for members of certain groups to gain admission to gateways courses, for example. After establishing what may be considered a "negative" effect of number-correct scoring and how this effect can be "corrected" by uses of certain kinds of formulas, the authors provide suggestions for using test directions and procedures as an alternative method of enhancing fairness. We are left wondering if, when all is said and done, it will ever be possible to use this kind of testing in ways that are fair, efficient, and still tell us what we want to know about those who took the test.

Thomas DeFranco is concerned with the dual nature of mathematics: as a body of knowledge and as a practice of solving problems. A first question concerns the difference between the way in which an expert and a novice might go about solving problems. But this paper points to pilot studies that suggest the question is not so simple. What is an expert? Does it matter whether the problems are susceptible to direct solution if one has the requisite knowledge vs. problems whose content may be elementary, but the method of solution far from obvious? DeFranco looks at two groups of Ph.D. mathematicians in which the only difference is that members of the first group had national or international reputations, whereas members of the second group did not. Both groups worked on a set of problems with elementary content, but whose solutions were not immediate. Unsurprisingly, the individuals in the first group outperformed those in the second by a long shot. What is interesting in this paper is the detailed analysis of the differences in strategies which presumably caused the differences in performance. The results have two implications. First, being a working mathematician does not guarantee strong problem-solving skills. Second, our graduate schools do not really reflect the two kinds of mathematical ability and the two sides of mathematics, but instead concentrate on knowledge rather than problem-solving skills.

Finally, maybe we are establishing a tradition. The editors have decided to run, as in *RCME I*, a list of questions related to Undergraduate Mathematics Education. In *RCME I*, Lynn Steen posed "Twenty Questions." In this issue we print a list of eighteen questions raised at the first Oberwolfach Conference in Undergraduate Mathematics Education, held in the Fall of 1995. These questions, relating to both research and curriculum, have no official status other than being the product of several hours' discussion by twenty-four people who decided that they were important. Perhaps we will maintain this tradition in future volumes. So we end the Preface with yet another open question: How many questions in *RCME III*?

Jim Kaput
Alan Schoenfeld
Ed Dubinsky

CBMS Issues in Mathematics Education
Volume **6**, 1996

A Framework for Research and Curriculum Development in Undergraduate Mathematics Education

MARK ASIALA, ANNE BROWN, DAVID J. DEVRIES,
ED DUBINSKY, DAVID MATHEWS, AND KAREN THOMAS

ABSTRACT. Over the past several years, a community of researchers has been using and refining a particular framework for research and curriculum development in undergraduate mathematics education. The purpose of this paper is to share the results of this work with the mathematics education community at large by describing the current version of the framework and giving some examples of its application.

Our framework utilizes qualitative methods for research and is based on a very specific theoretical perspective that is being developed through attempts to understand the ideas of Piaget concerning reflective abstraction and reconstruct them in the context of college level mathematics. Our approach has three components. It begins with an initial theoretical analysis of what it means to understand a concept and how that understanding can be constructed by the learner. This leads to the design of an instructional treatment that focuses directly on trying to get students to make the constructions called for by the analysis. Implementation of instruction leads to the gathering of data, which is then analyzed in the context of the theoretical perspective. The researchers cycle through the three components and refine both the theory and the instructional treatments as needed.

In this report the authors present detailed descriptions of each of these components. In our discussion of theoretical analyses, we describe certain mental constructions for learning mathematics, including actions, processes, objects, and schemas, and the relationships among these constructions. Under instructional treatment, we describe the components of the ACE teaching cycle (activities, class discussion, and exercises), cooperative learning and the use of a mathematical programming language. Finally, we describe the methodology used in data collection and analysis. The paper concludes with a discussion of issues raised in the use of this framework, followed by an extensive bibliography.

1. Introduction

The purpose of this paper is to set down a very specific methodology for research in the learning of mathematics, and curriculum development based on that research. We are concerned with theoretical analyses which model mathematical understanding, instruction based on the results of these analyses, and empirical data, both quantitative and qualitative, that can be used to refine the theoretical perspective and assess the effects of the instruction.

This report caps several years of research and development during which a framework for conducting this work has been developed and applied to various topics in collegiate mathematics. Now that the main outlines of our approach have stabilized, and the group of researchers using this framework has begun to grow, we wish to share our work with others in the mathematics education community, including potential collaborators.[1] We offer here the details of our approach, both as a report on what has taken place and as a possible guide to future efforts.

After a few general remarks in the introduction about paradigms, and about previous discussions of the specific framework we have been developing, we proceed to an overview of our approach as it now stands, and the goals we associate with it. Then, in the main portion of the paper, we describe in detail our framework and its components along with several examples. Next, we consider some larger issues that arise in connection with our framework. Finally, we summarize what has been said in this paper and point to what can be expected in future reports.

1.1 What is a paradigm and why is one needed?

The seminal work of Thomas Kuhn [19] teaches us that scientific research proceeds according to what he calls *paradigms*.[2] *A paradigm is a collection of understandings (explicit or implicit) on the part of an individual or group of individuals about the kinds of things one does when conducting research in a particular field, the types of questions that are to be asked, the sorts of answers that are to be expected, and the methods that are to be employed in searching for these answers.*

We also learn from Kuhn that "paradigm shifts" do not come quickly or easily, but they do tend to be sharp. They are caused by what Kuhn calls a "crisis state" which can be the result of one of two different situations. One is that an event or discovery is so far reaching that it is impossible to assimilate it into the current paradigm, hence the need for a new paradigm. The second situation is a developing dissatisfaction with the current paradigm that reaches

---

[1] We have organized an informal community of researchers who use this framework, known as *Research in Undergraduate Mathematics Education Community* or *RUMEC*, with the intention of producing a series of research reports by various subsets of the membership. The present paper is the first in the *RUMEC* series.

[2] For a critical look at Kuhn's and others' philosophical views on the structure of scientific revolutions and theories see Suppe [30] or Gutting [16].

a level where answers to certain questions (often basic to the field) can not be easily or satisfactorily obtained. This brings about a sort of declared rejection of the current paradigm by individuals who begin to search for a new paradigm. Whatever the cause, the crisis state continues until an alternative can be agreed upon by the community and then the change *must* be a simultaneous rejection of the old paradigm and an acceptance of the new paradigm by the research community as a whole. Thus, dissatisfaction with a particular paradigm builds up gradually over a long period of time until there occurs a moment, or "scientific revolution" in which the understandings change rather drastically and fairly quickly.

It seems that the second situation which causes Kuhn's scientific revolution has occurred for research in mathematics education. For a long time, research in this field consisted almost exclusively of statistical comparisons of control and experimental populations according to designs proposed by Sir R. A. Fisher some 60 years ago for the purpose of making decisions about agricultural activities [12]. In the last decade or so, there has been a growing concern with the impossibility of really meeting the conditions required to make application of statistical tests to mathematics education valid, a dissatisfaction with the small differences and unrealistic contexts to which these designs seem to lead, and a developing understanding that the fundamental mechanisms of learning mathematics are not as simplifiable and controllable as agricultural factors. Traditional statistical measures may apply, for example to paired-associate learning, but if one wishes to build on the work of Jean Piaget, and/or use the ideas of theoretical cognitive structures, then new methods of research, mainly qualitative, must be developed to relate those structures to observable behavior. (For a discussion of the implications of cognitive science on research methodology in education see Davis [4].) Workers in the field have stopped insisting on a statistical paradigm and have begun to think about alternatives.

This represents only the conditions for a scientific revolution and is not yet a paradigm shift because no single alternative point of view has been adopted to replace the accepted paradigm. The ideas of Jean Piaget have influenced many researchers to turn from quantitative to qualitative methods, but there are many forms of qualitative research that are being used at present. In considering the variety of approaches being used, there are two aspects which must be addressed. The first is the theoretical perspective taken by the researchers using a particular approach, and the second is the set of actual methods by which data is collected and analyzed.

Patton [20] lists some of the theoretical perspectives used by qualitative researchers. These include the ethnographic perspective, in which the central goal is to describe the culture of a group of people, and related perspectives including phenomenology and heuristics, in which the goals are to describe the essential features of a particular experience for a particular person or group of people. The tradition of ecological psychology seeks to understand the effect of the setting on the ways in which people behave, and the perspective of systems theory seeks to describe how a particular system (for instance, a teacher and a group of students in a given classroom) functions. In addition, there are orientational approaches (feminist, Marxist, Freudian, etc.) within each of these perspectives.

Methods of data collection for qualitative studies vary widely, and in many of the theoretical perspectives it is considered important to use a wide variety of data collection methods, studying the phenomena of interest from all available angles in order to be able to triangulate data from many sources in reaching a conclusion. Romberg [24] discusses the use of interviews, which may range from informal discussions between researchers and participants to very structured conversations in which a predetermined list of questions is asked of each participant. He also discusses observational methods ranging from videotaping to the use of trained observers to participant observations. Patton [20] discusses other sources of data, including documents and files which may be available to the researcher, photographs and diagrams of the setting in which the research takes place, and the researchers' field notes.[3]

Thus we see that there is a wide variety of frameworks in which it is possible to work. In this paper we describe our choice which has been been made consciously with concern for the theoretical and empirical aspects as well as applicability to real classrooms in the form of instructional treatments.

According to Kuhn, researchers shift to a new framework because it satisfies the needs of the times more than the existing paradigm. There are two reasons why we feel that researchers should make conscious choices about the framework under which they work.

One reason is necessity. The variety of qualitative research methodologies indicated above does not appear to be leading to any kind of convergence to a single approach (or even a small number of approaches) that has general acceptance. We are finding more and more that research done according to one framework is evaluated according to another and this is leading to some measure of confusion. Therefore, we feel that researchers should make explicit the framework they are using and the basis on which their work is to be judged. Consumers of the results of research need to have a clear idea of what they can and cannot expect to get out of a piece of research.

Second, we feel that a conscious attention to the specifics of one's framework is more in keeping with the scientific method as expressed by David Griese who interpreted science as "a department of practical work which depends on the knowledge and conscious application of principle" [15]. Griese decided 20 years ago that it was time to move computer programming to an endeavor in which it was possible to teach the principles so that they can be consciously applied. We believe that it is important today that those who study the learning of post-

---

[3]For information on qualitative methods used in the social sciences see Jacob [17] or Patton [20]. For information on qualitative methods used in mathematics education research see Romberg [24] or Schoenfeld [26].

secondary mathematics attempt to make available to others the methodologies under which they work.

1.2 Previous discussions of this framework.

Components of our framework have been discussed in several papers over the last several years: [2,5,7,8]. The overall framework with its three components have been discussed at length, especially the theoretical component, but only very fragmented discussions of the other two components, instructional treatments and gathering/analyzing data have been given. Moreover, the framework and its components are continuing to evolve as we reflect on our practice. Finally, the authors of this article are part of a larger community which is in the process of producing a number of studies of topics in calculus and abstract algebra using this framework. Therefore, it seems reasonable at this time to present a complete, self-contained and up-to-date discussion of the entire framework.

2. A framework for research and curriculum development

2.1 Overview of the framework.

The framework used in this research consists of three components. Figure 1 illustrates each of these components and the relationships among them. A study of the cognitive growth of an individual trying to learn a particular mathematical concept takes place by successive refinements as the investigator repeatedly cycles through the component activities of Figure 1.

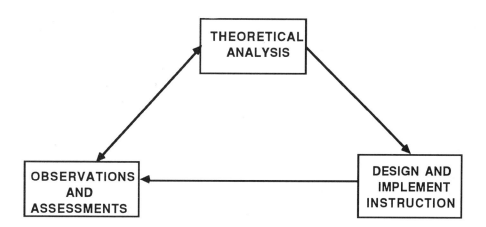

FIGURE 1. The framework.

Research begins with a theoretical analysis modeling the epistemology of the concept in question: what it means to understand the concept and how that understanding can be constructed by a learner. This initial analysis, marking the researchers' entry into the cycle of components of the framework, is based primarily on the researchers' understanding of the concept in question and on their experiences as learners and teachers of the concept. The analysis informs the design of instruction. Implementing the instruction provides an opportunity for gathering data and for reconsidering the initial theoretical analysis with respect to this data. The result may well be a revision of the theoretical analysis which then lays the foundation for the next iteration of the study. This next iteration begins with the revised theoretical analysis and ends with a further revision or deeper understanding of the epistemology of the concept in question which may become the foundation for yet another repetition of the cycle. These repetitions are continued for as long as appears to be necessary to achieve stability in the researchers' understanding of the epistemology of the concept.

2.2 Goals and issues associated with the framework.

Research using this framework is inevitably a synthesis of "pure" and "applied" research. Each time the researchers cycle through the components of the framework, every component is reconsidered and, possibly, revised. In this sense the research builds on and is dependent upon previous implementations of the framework. We observe students trying to understand mathematics and offer explanations of successes and failures in terms of mental constructs and the ways in which they transform. Our specific goals are: to increase our understanding of how learning mathematics can take place, to develop a theory-based pedagogy for use in undergraduate mathematics instruction, and to develop a base of information and assessment techniques which shed light on the epistemology and pedagogy associated with particular concepts. The goals are thus associated with the three components of the framework.

There are many issues raised as a result of the use of this framework. In the component of theoretical analysis there are the following issues: (1) How does one go about developing the theoretical perspective? (2) How do we see the relationship between this theory and what actually happens; that is, to what extent can a theoretical analysis provide an accurate or even approximate picture of what is going on in the minds of the learners? In the component dealing with pedagogy there is the issue of explaining the relationship between the instructional treatments and our theoretical analysis. With respect to data analysis there are the following issues: (1) To what extent do our theoretical ideas work? (2) How much mathematics is being learned by the students? (3) What would it take to falsify specific conjectures or our theory in general? (4) Since data can come from this study but also from assessment of student learning which may not be part of this study, what is the appropriate use of these in drawing conclusions?

We will return to these issues later in this paper in Section 4, but in order to do so, it is important to develop more fully the description of the three components of the framework and their interconnections (as illustrated by the arrows in Figure 1).

3. The components of the framework

3.1 Theoretical analysis.

The purpose of the theoretical analysis of a concept is to propose a model of cognition: that is, a description of specific mental constructions that a learner might make in order to develop her or his understanding of the concept. We will refer to the result of this analysis as a *genetic decomposition* of the concept. That is, a genetic decomposition of a concept is a structured set of mental constructs which might describe how the concept can develop in the mind of an individual.

The analysis is initially made by applying a general theory of learning and is greatly influenced by the researchers' own understanding of the concept and previous experience in learning and teaching it. In subsequent iterations through the framework, the analysis of data increasingly contributes to the evolving genetic decomposition.

In working with this framework we make use of a very specific theoretical perspective on learning which has developed through our attempt to understand the ideas of Piaget concerning reflective abstraction and to reconstruct these ideas in the context of college-level mathematics. The initial development of this theory and its relationship to Piaget is described in some detail in [6]. The perspective is continuing to develop and we describe it here in its present form. We should note that, although our theoretical perspective is closely related to the theories of Piaget, this is not so much the case for the other components of our framework. Indeed, considerations of pedagogical strategies are almost absent from the totality of Piaget's work and our methodology for gathering and analyzing data is influenced in only some, but not all, of its aspects by the methodology which Piaget used.

3.1.1 Mathematical knowledge and its construction.

Our theoretical perspective begins with a statement of our overall perspective on what it means to learn and know something in mathematics. The following paragraph is not a definition, but rather an attempt to collect the essential ingredients of our perspective in one place.

> An individual's mathematical knowledge is her or his tendency to respond to perceived mathematical problem situations by reflecting on problems and their solutions in a social context and by constructing or reconstructing mathematical actions, processes and objects and organizing these in schemas to use in dealing with the situations.

There are, in this statement, references to a number of aspects of learning and knowing. For one thing, the statement acknowledges that what a person knows and is capable of doing is not necessarily available to her or him at a given moment and in a given situation. All of us who have taught (or studied) are familiar with the phenomenon of a student missing a question completely on an exam and then really knowing the answer right after, without looking it up. A related phenomenon is to be unable to deal with a mathematical situation but, after the slightest suggestion from a colleague or teacher, it all comes running back to your consciousness. Thus, in the problem of knowing, there are two issues: learning a concept and accessing it when needed.

Reflection, in the sense of paying conscious attention to operations that are performed, is an important part of both learning and knowing. Mathematics in particular is full of techniques and algorithms to use in dealing with situations. Many people can learn these quite well and use them to do things in mathematics. But understanding mathematics goes beyond the ability to perform calculations, no matter how sophisticated. It is necessary to be aware of how procedures work, to get a feel for the result without actually performing all of the calculations, to be able to work with variations of a single algorithm, to see relationships and to be able to organize experiences (both mathematical and non-mathematical).

From this perspective we take the position that reflection is significantly enhanced in a social context. There is evidence in the literature (see [32], for example) for the value to students of social interaction and there is also the cultural reality that virtually all research mathematicians feel very strongly the need for interactions with colleagues before, during, and after creative work in mathematics.

The statement describing our theoretical perspective asserts that "possessing" knowledge consists in a tendency to make mental constructions that are used in dealing with a problem situation. Often the construction amounts to reconstructing (or remembering) something previously built so as to repeat a previous method. But progress in the development of mathematical knowledge comes from making a reconstruction in a situation similar to, but different in important ways from, a problem previously dealt with. Then the reconstruction is not exactly the same as what existed previously, and may in fact contain one or more advances to a more sophisticated level. This whole notion is related to the well known Piagetian dichotomy of assimilation and accommodation [21]. The theoretical perspective which we are describing is itself the result of reconstruction of our understanding of Piaget's theory leading to extension in its applicability to post-secondary mathematics.

Finally, the question arises of what is it that is constructed by the learner, or, in other words, what is the nature of the constructions and the ways in which they are made? As we turn to this issue, it should become apparent that our theoretical perspective, which may appear applicable to any subject whatsoever, becomes specific to mathematics.

3.1.2 Mental constructions for learning mathematics.

As illustrated in Figure 2, we consider that understanding a mathematical concept begins with manipulating previously constructed mental or physical objects to form actions; actions are then interiorized to form processes which are then encapsulated to form objects. Objects can be de-encapsulated back to the processes from which they were formed. Finally, actions, processes and objects can be organized in schemas. A more detailed description of each of these mental

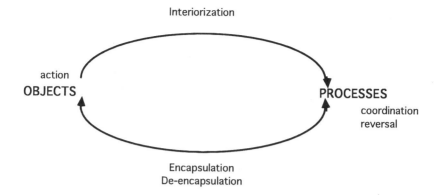

FIGURE 2. Constructions for mathematical knowledge.

constructions is given below.

In reading our discussion of these specific constructions, we would like the reader to keep in mind that each of the four is based on a specific construction of Piaget, although not always with exactly the same name. What we are calling actions are close to Piaget's *action schemes*, our processes are related to his *operations*, and *object* is one of the terms Piaget uses for that to which actions and processes can be applied. The term *schema* is more difficult, partly because Piaget uses several different terms in different places, but with very similar meanings, partly because of the difficulty of translating between two languages which both have many terms related to schema with subtle and not always corresponding distinction, and partly because our understanding of this specific construction is not as far as advanced as it is for the others. Our use of the term in this paper is close to what Piaget calls *schemata* in [22] where his meaning appears to be in some ways similar to the *concept image* of Tall and Vinner [31], and where he talks about thematizing schemas which, essentially refers to making them objects. For a more detailed discussion of the relation between the theories of Piaget and (an early version of) our theoretical framework, the reader is referred to [6].

Action. An action is a transformation of objects which is perceived by the individual as being at least somewhat external. That is, an individual whose understanding of a transformation is limited to an action conception can carry out the transformation only by reacting to external cues that give precise details on what steps to take.

For example, a student who is unable to interpret a situation as a function unless he or she has a (single) formula for computing values is restricted to an action concept of function. In such a case, the student is unable to do very much with this function except to evaluate it at specific points and to manipulate the formula. Functions with split domains, inverses of functions, composition of functions, sets of functions, the notion that the derivative of a function is a function, and the idea that a solution of a differential equation is a function are all sources of great difficulty for students. According to our theoretical perspective, a major reason for the difficulty is that the learner is not able to go beyond an action conception of function and all of these notions require process and/or object conceptions. (See [1] for an elaboration of these issues.)

Another example of an action conception comes from the notion of a (left or right) coset of a group in abstract algebra. Consider, for example, the modular group $[\mathcal{Z}_{20}, +_{20}]$ — that is, the integers $\{0, 1, 2 \ldots, 19\}$ with the operation of addition mod 20 — and the subgroup $H = \{0, 4, 8, 12, 16\}$ of multiples of 4. As is seen in [10] it is not very difficult for learners to work with a coset such as $2 + H = \{2, 6, 10, 14, 18\}$ because it is formed either by an explicit listing of the elements obtained by adding 2 to each element of H or applying some rule (e.g., "begin with 2 and add 4") or an explicit condition such as, "the remainder on division by 4 is 2". Understanding a coset as a set of calculations that are actually performed to obtain a definite set is an action conception. Something more is required to work with cosets in a group such as $\mathcal{S}_n$, the group of all permutations on n objects where simple formulas are not available. Even in the more elementary situation of $\mathcal{Z}_n$, students who have no more than an action conception will have difficulty in reasoning about cosets (such as counting them, comparing them, etc.). In the context of our theoretical perspective, these difficulties are related to a student's inability to interiorize these actions to processes, or encapsulate the processes to objects.

Although an action conception is very limited, the following paragraphs describe the way in which actions form the crucial beginning of understanding a concept. Therefore, our learning-theory-based pedagogical approach begins with activities designed to help students construct actions.

Process. When an action is repeated, and the individual reflects upon it, it may be interiorized into a process. That is, an internal construction is made that performs the same action, but now, not necessarily directed by external stimuli. An individual who has a process conception of a transformation can reflect on, describe, or even reverse the steps of the transformation without actually performing those steps. In contrast to an action, a process is perceived by the individual as being internal, and under one's control, rather than as something one does in response to external cues.

In the case of functions, a process conception allows the subject to think of a function as receiving one or more inputs, or values of independent variables, performing one or more operations on the inputs and returning the results as outputs, or values of dependent variables. For instance, to understand a function such as $\sin x$, one needs a process conception of function since no explicit instructions for obtaining an output from an input are given; in order to implement the function, one must imagine the process of associating a real number with its sine.

With a process conception of function, an individual can link two or more processes to construct a composition, or reverse the process to obtain inverse functions [1].

In abstract algebra, a process understanding of cosets includes thinking about the formation of a set by operating a fixed element with every element in a particular subgroup. Again, it is not necessary to perform the operations, but only to think about them being performed. Thus, with a process conception, cosets can be formed in situations where formulas are not available. (See, for example, [10].)

Object. When an individual reflects on operations applied to a particular process, becomes aware of the process as a totality, realizes that transformations (whether they be actions or processes) can act on it, and is able to actually construct such transformations, then he or she is thinking of this process as an object. In this case, we say that the process has been *encapsulated* to an object.

In the course of performing an action or process on an object, it is often necessary to de-encapsulate the object back to the process from which it came in order to use its properties in manipulating it.

It is easy to see how encapsulation of processes to objects and de-encapsulating the objects back to processes arises when one is thinking about manipulations of functions such as adding, multiplying, or just forming sets of functions. In an abstract algebra context, given an element x and a subgroup H of a group G, if an individual thinks generally of the (left) coset of x modulo H as a process of operating with x on each element of H, then this process can be encapsulated to an object xH. Then, cosets are named, operations can be performed on them [10], and various actions on cosets of H, such as counting their number, comparing their cardinality, and checking their intersections can make sense to the individual. Thinking about the problem of investigating such properties involves the interpretation of cosets as objects whereas the actual finding out requires that these objects be de-encapsulated in the individual's mind so as to make use of the properties of the processes from which these objects came (certain kinds of set formation in this case).

In general, encapsulating processes to become objects is considered to be extremely difficult [22,28,29] and not very many pedagogical strategies have been effective in helping students do this in situations such as functions or cosets. A part of the reason for this ineffectiveness is that very little (if anything) in our experience corresponds to performing actions on what are interpreted as processes.

Schema. Once constructed, objects and processes can be interconnected in various ways: for example, two or more processes may be coordinated by linking them (through composition or in other ways); processes and objects are related by virtue of the fact that the former act on the latter. A collection of processes and objects can be organized in a structured manner to form a schema. Schemas themselves can be treated as objects and included in the organization of "higher level" schemas. When this happens, we say that the schema has been *thematized* to an object. The schema can then be included in higher level schemas of mathematical structures. For example, functions can be formed into sets, operations on these sets can be introduced, and properties of the operations can be checked. All of this can be organized to construct a schema for function space which can be applied to concepts such as dual spaces, spaces of linear mappings, and function algebras.

As we indicated above, our work with, and understanding of, the idea of a schema has not progressed as much as some of the other aspects of our general theory. We are convinced, however, that this notion is an important part of the total picture and we hope to understand it better as our work proceeds. For now, we can only make some tentative observations concerning, for example, the distinction between a genetic decomposition and a schema.

A related question, on which our work is also very preliminary, has to do with relations between the mental constructs together with the interconnections that an individual uses to understand a concept, and the way in which an individual uses (or fails to use) them in problem situations. This is what we were referring to in our use of the term "tendency" in our initial statement. Now, a genetic decomposition for a mathematical concept is a model used by researchers to describe the concept. Our tentative understanding suggests that an individual's schema for a concept includes her or his version of the concept that is described by the genetic decomposition, as well as other concepts that are perceived to be linked to the concept in the context of problem situations.

Put another way, we might suggest that the distinction between schema and other mental constructions is like the distinction in biology between an organ and a cell. They are both objects, but the organ (schema) provides the organization necessary for the functioning of the cells to the benefit of the organism. An individual's schema is the totality of knowledge which for her or him is connected (consciously or subconsciously) to a particular mathematical topic. An individual will have a function schema, a derivative schema, a group schema, etc. Schemas are important to the individual for mathematical empowerment, but in general, we are very far from knowing all of the specifics, nor have we studied much about how this organization determines mathematical performance. All we can do now is to link together mental constructions for a concept in a generic road-map of development and understanding (genetic decomposition) and see how this is actualized in a given individual (schema). Clearly, an individual's schema may include actions or knee-jerk responses such as, "whenever I see this symbol I do that."

3.2 Instructional treatments.

The second component of our framework has to do with designing and implementing instruction based on the theoretical analyses. The theoretical perspective on learning we have just described influences instruction in two ways. First, as we have indicated earlier, the theoretical analysis postulates certain specific mental constructions which the instruction should foster. Later in the paper (Section 3.2.2) we will consider that effect in relation to specific mathematical content. Before doing that however, we shall describe a second, more global way in which the general theory influences instruction.

3.2.1 Global influences of the theoretical perspective.

Returning to our formulation of the nature of mathematical knowledge and its acquisition (Section 3.1.1) we consider four components: the tendency to use one or another mathematical construct, reflection, social context, and constructions or reconstructions.

When confronted with a mathematical problem situation, it has frequently been observed that an individual is not always able to bring to bear specific ideas in her or his mathematical repertoire. (See, for instance, [25] for a very sharp example of students clearly possessing certain knowledge, but not being able to use it.) An individual's mathematical knowledge consists in a tendency to use certain constructions, but not all relevant constructions are recalled in every situation. This is one of the reasons that we cannot expect students to learn mathematics in the logical order in which it can be laid out. In fact, according to our theoretical perspective, the growth of understanding is highly non-linear with starts and stops; the student develops partial understandings, repeatedly returns to the same piece of knowledge, and periodically summarizes and ties related ideas together [27]. Our general instructional approach acknowledges this growth pattern by using what we call an *holistic spray*, which is a variation on the standard spiral method [10].

In this variation, students are thrust into an intentionally disequilibrating environment which contains as much as possible about the material being studied. The idea is that everything is sprayed at them in an holistic manner, as opposed to being sequentially organized. Each individual (or cooperative learning group) tries to make sense out of the situation—that is, they try to solve problems, answer questions or understand ideas. Different students may learn different pieces of the whole at different times. In this way the students enhance their understanding of one or another portion of the material bit by bit. The course keeps presenting versions of the whole set of material and the students are always trying to make more sense, always learning a little more.

The social context to which our theoretical perspective refers is implemented in our instruction through the use of cooperative learning groups. Students are organized at the beginning of the semester in small groups of three to five to do all of the course work (computer lab, class discussion, homework and some exams) cooperatively. For details see [23]. One consequence of having students work in cooperative groups is that they are more likely to reflect on the procedures that they perform [32].

A critical part of our approach is to implement the results of our theoretical analysis in regular classrooms and to gather data on what happened in those classrooms. Consequently, our curricular designs are tied to the current curricular structures of the semester or quarter system in which we operate. Today, many people are thinking seriously about alternative ways to organize the entire educational enterprise—from kindergarten through graduate school. As new structures emerge, the generality of our approach will be indicated by the extent to which it can be adapted to these new forms.

Our instructional strategies also try to get students to reflect on their work through the overall structure of a course. A particular pedagogical approach we use, which we refer to as the *ACE Teaching Cycle*, and which we now describe, is not a necessary consequence of the theoretical perspective but is one possible overall design supporting our theoretical analyses. In this design, the course is broken up into sections, each of which runs for one week. During the week, the class meets on some days in the computer lab and on other days in a regular classroom in which there are no computers. Homework is completed outside of class. As indicated above, the students are in cooperative groups for all of this work.

Thus there are three components of the ACE cycle: activities, class discussion, and exercises.

Activities. Class meets in a computer lab where students work in teams on computer programming tasks designed to foster specific mental constructions suggested by the theoretical analysis. The lab assignments are generally too long to finish during the scheduled lab and students are expected to come to the lab when it is open, or work on their personal computers, or use other labs to complete the assignment. It is important to note that there are major differences between these computer activities and the kinds of activities used in "discovery learning." While some computer activities may involve an element of discovery, their primary goal is to provide students with an experience base rather than to lead them to correct answers. Through these activities students gain experience with the mathematical issues which are later developed in the classroom phase. We will discuss the computer work in more detail below.

Class Discussion. Class meets in a classroom where students again work in teams to perform paper and pencil tasks based on the computer activities in the lab. The instructor leads inter-group discussions designed to give students an opportunity to reflect on the work they did in the lab and calculations they have been making in class. On occasion, the instructor will provide definitions, explanations and overviews to tie together what the students have been thinking about.

Exercises. Relatively traditional exercises are assigned for students to work on in teams. These are expected to be completed outside of class and lab and they represent homework that is in addition to the lab assignments. The purpose of the exercises is for students to reinforce the ideas they have constructed, to use the mathematics they have learned and, on occasion, to begin thinking about situations that will be studied later.

One must consider the question of how our pedagogical strategies are supported by the textbook for the course. We have found that traditional textbooks are not very helpful. For example, template problems (usually a key feature of a textbook) circumvent the disequilibration and formation of rich mental constructions which we consider necessary for meaningful understanding. Also, as we have indicated, mathematics is not learned in the logical order in which it is presented in most textbooks. It has therefore been necessary to produce appropriate texts for various courses. This has been done, using essentially the same style, for courses in Discrete Mathematics, Precalculus, Calculus, and Abstract Algebra. Other works are in progress. The textbooks are designed to support the constructivist approach to teaching that is described in this paper.

Our textbooks are arranged according to the ACE Cycle. Each book is divided into sections, each of which begins with a set of computer activities, followed by a discussion of the mathematics involved in these activities, and ending with an exercise set. It is expected that a section will be covered in one week with the students doing the computer activities in the lab, and possibly during "open hours," working through the discussion material in class and doing the exercises as homework.

The textbooks have certain features that, although not always popular with students *or* teachers, we feel are necessary to relate to how learning actually takes place. There are almost no cases of "worked problems" followed by drills in the exercises and no answers to exercises in the back of the book. Wherever possible, the student is given an opportunity to figure out for her or himself how to solve a problem using the ideas that are being learned. In the computer activities, the students are often asked to solve problems requiring mathematics they have not yet studied. They are encouraged to discuss these issues with their group members or other colleagues, read ahead in the text (where explanations can be found, although usually not directly), or even consult other texts. In other words, the students are forced to *investigate* mathematical ideas in order to solve problems. The explanations that are given in the discussion section are interspersed with many questions whose answers are important for understanding the concepts. These questions are not always answered in the texts (at least not right away) but they are generally repeated in the exercise set.

Although we feel that our textbooks make a major contribution to student learning, the books we produce are not as good for later reference. To overcome this, we try to include an extremely detailed and complete index, but also expect that the student might find some other text which is not so helpful for learning, but may be a better reference book.

3.2.2 Mental constructions.

Our main strategy for getting students to make mental constructions proposed by our theoretical analysis is to assign tasks that will require them to write and/or revise computer code using a mathematical programming language. In the sequel, we will refer to this as an *MPL*. The idea is that when you do something on a computer, it affects your mind. This effect occurs in three ways.

First, we attempt to orchestrate the effect of these tasks so that we foster the specific mental constructions proposed by our theoretical analyses. Second, the computer tasks which we assign can provide a concrete experience base paving the way for later abstraction. Finally, there is an indirect effect from working with computers which has been reported and is less clearly understood.

Fostering mental constructions directly. This general discussion of our strategy is organized around actions, processes, objects, and schemas. The initial activities included in this section show how students familiarize themselves with the syntax of the programming language. At the same time, since the activities involve direct calculations of specific values, students gain experience constructing actions corresponding to selected mathematical concepts. This experience is built upon in subsequent activities where students are asked to reconstruct familiar actions as general processes. Later activities presented exemplify those that are intended to help students encapsulate processes to objects; these activities typically involve writing programs in which the processes to be encapsulated are inputs and/or outputs to the program. Finally, we describe a more complicated activity in which students need to organize a variety of previously constructed objects into a schema that can then be applied to particular problem situations.

Although our choice for a language in which to work is *ISETL*, there are other possibilities such as *Mathematica* and *Maple*. Because the syntax of *ISETL* is so close to that of standard mathematical notation, the programming aspects of the language are particularly easy to learn. This does not mean, however, that learning to write correct mathematics in *ISETL* is easy. On the contrary, students encounter a great many difficulties in using the language—difficulties that are most often directly associated with mathematical difficulties. For this reason, *ISETL* provides an ideal environment for mathematical experimentation, reflection, and discussion.

For those who know mathematics and have some experience with programming, little explanation is necessary in describing examples of the use of *ISETL*. Hence we will minimize our explanations of syntax in the sequel. (See [8] for a discussion of the use of this language and [3] for details on its syntax.)

Actions. We begin with activities in which students try to repeat on their terminal screens what is written in the text, or to predict what will be the result of running code that is given to them, or to modify code they have been given.

Following is a set of *ISETL* instructions as they would appear on the screen followed by the computer's response. It is taken from the first pages of the textbook used in the C^4L calculus project [11].

The > symbol is the *ISETL* prompt and lines which begin with this symbol or >>, which indicates incomplete input, are entered by the user. Lines without these prompts are what the computer prints on the screen.

```
>    7+18;
25;


>    13*(233.8);
3039.400000;


>    5 = 2.0 + 3;
true;


>    4 >= 2 + 3;
false;
>    17
>>   + 23.7 - 46
>>   *2
>>   ;
-51.300000;
>    x := -23/27;
>    x;
-0.851852;


>    27/36;
0.750000;


>    p := [3,-2]; q := [1,4.5]; r := [0.5,-2,-3];
>    p; q; r;
[3, -2];
[1, 4.500000];
[.500000, -2, -3];
>    p(1); p(2); q(2); r(3);
3;
-2;
4.500000;
-3;
>    p(1)*q(1) + p(2)*q(2);
-6.000000;


>    length := 0;
>    for i in [1..3] do
>>       length := length + r(i)**2;
>>       end;
>    length := sqrt(length);
>    length;
3.640055;
```

Students are asked to do these exercises for the purpose of becoming familiar with their *MPL*, but at the same time, there are several mathematical concepts which they have an opportunity to construct at the action level. For example, there are simple propositions, the formation of pairs and triples of numbers and the action of picking out an indexed term of a given sequence. Also the concept of dot product appears as an action. Finally, the algorithm for computing the length of a vector in three dimensions appears as an action because the calculation is explicit and is applied to a single vector.

Processes. In the previous paragraph we indicated how students might be led to construct actions by writing code that makes a computation once with specific numbers. Now we try to get them to build on these actions by performing activities intended to help them reconstruct their actions as processes. For example, students are led to interiorize actions to processes when they replace code written to perform a specific calculation by a computer function which will perform the action for any values. Thus, for the calculation of the length of a three-dimensional vector (which is our last example above), we might ask students to write the following computer function. (The last two lines assume that r has been defined previously as above, or as any three-dimensional vector.)

```
>    length := func(v);
>>       l := 0;
>>       for i in [1..3] do
>>            l := l + v(i)**2;
>>            end;
>>       return sqrt(l);
>>   end;
>    length(r);
3.64005;
```

There are a variety of ways that programming activities can be used to help students to understand functions as processes. Research suggests that students who write code to implement the point-wise sum, product, and composition of functions tend to make progress in developing a process conception of function [1]. Programming activities have also been used to help students get past the well-known difficulty many have in seeing that a piece-wise defined function is still a function, and that properties can be studied at the seam points as well as other places. A process conception of function emerges as students write programs in which the definition of a piece-wise function is implemented through the simple use of a conditional:

```
f :=func(x);
        if x <= 1 then return 2-x**2;
        else return x/2 + 1.5;
        end;
    end;
```

Following is how the code would look in *Maple*.

```
f:=proc(x);
if x<=1 then 2-x**2;
else x/2 + 1.5;
fi;
end;
```

A completely different approach considered by Kaput has students (in grades 3-13) *begin* with piecewise-defined functions and deal with them graphically using what is referred to as SimCalc simulations. (See [18] for details.)

We can also mention the example of boolean-valued functions of the positive integers. Our research suggests that one of the difficulties students have with proof by induction is at the very beginning. A student is faced with a problem: show that a certain statement involving an arbitrary integer is true for all (sufficiently large) values of the integer. This kind of problem is very new and difficult for most students. It really is a (mental) function which accepts a positive integer and plugs it into the statement to obtain a proposition which may be true or false — and the answer could be different for different values of n. Once again, expressing this problem as a function in an *MPL* is a big help for students in figuring out how to begin.

Suppose, for example, that the problem is to determine if a gambling casino with only $300 and $500 chips can represent (within the nearest $100) any amount of money beyond a certain minimum. We encourage students to begin their investigation by writing a computer program that accepts a positive integer and returns a boolean value. The following is one solution they generally come up with in our elementary discrete mathematics course.

```
P :=func(n);
        if is_integer(n)
            and n > 0
            and exists x,y in [0..n div 3] | 3*x + 5*y = n
        then return true;
        else return false;
        end;
    end;
```

Objects. Objects are obtained by encapsulation of processes, and an individual is likely to do this when he or she reflects on a situation in which it is necessary to apply an action to a dynamic process. This presents a difficulty because the action cannot be applied to the process until after the process has been encapsulated to an object. But, as we have said before, mental constructions do not seem to occur in simple logical sequences. In fact, the following three things can all be happening at the same time, initially in some amorphous combination:

1. the need to create an object (in order to apply an action to a process),
2. the encapsulation of the process to form the object, and
3. the application of an action to that object.

Gradually, as the learner reflects, he or she is able to differentiate, reorganize, and integrate the components of this experience so that a clear application of the action to the object is apparent.

Consider, for example, a student who has just learned what a permutation of $\{1, \ldots, n\}$ is, and is now faced with composing permutations. On the one hand, he or she could focus on permutations as processes, and simply perform the linking of the processes to get the composition. That would require only a process, but not an object conception of permutation. At a higher level, though, if the student is trying to think of composition as a binary operation, he or she would begin to see permutations as inputs to the binary operation process, and thus as objects.

Getting students to do all of this is another matter and there are very few effective pedagogical methods known. As we have indicated, one such method is to put students in situations where a problem must be solved or a task must be performed by writing programs in which the processes to be encapsulated are inputs to, and/or outputs of, the programs. Thus, in addition to having syntax similar to standard mathematical notation, we require that an *MPL* treat functions as first-class data types, that is, as entities which can be passed as input or output parameters to and from other student-defined procedures.

Continuing with our treatment of mathematical induction, we can report that students learn to treat propositions about natural numbers as objects. At the same time they develop an understanding of the "implication from n to $n + 1$", that is, $P(n) \longrightarrow P(n+1)$ understood as an object whose truth value as n varies is to be considered. Our approach is to have them write and apply the following program which accepts a function whose domain is the positive integers and whose range is the two element set $\{\texttt{true}, \texttt{false}\}$. This program returns the corresponding implication-valued function. (The symbol $ refers to a comment and anything after this symbol on the line in which it appears is ignored by *ISETL*.)

```
implfn :=    func(P); $ P is a boolean-valued function.
                return func(n);
                     return P(n) impl P(n+1);
                end;
          end;
```

In calculus, two extremely important examples of construction of objects occur in connection with derivatives and integrals. Although it is very simple for mathematicians, our experience suggests that the idea that the derivative of a function is a function is not immediate for students. Writing a program such as the following, which accepts a function and returns an approximation to its derivative appears to help.

```
df  :=    func(f);
              return fun c(x);
                      return (f(x + 0.00001) - f(x))/0.00001;
                  end;
          end;
```

In this program, f is the variable and so it does not need to be defined before the program is run. Once it has been run, a function can be defined and given any name, for example g and then df(g) will be a function that can be evaluated, assigned to be the value of a variable, graphed or treated in any other way that functions are treated.

It is important to note that the reason we are using *ISETL* here is because there are essential understandings we are trying to get students to construct as specified by our genetic decomposition, and we cannot do this with many other systems. For example, writing a program that constructs a function for performing a specific action in various contexts tends to get students to interiorize that action to a process. Perhaps even more critical is our method of getting students to make a function an object in their minds by using function programs as input to another program which students write. This latter program performs a process and returns a function as output. This can also be done using systems such as *Maple* or *Mathematica* as alternatives to *ISETL*. For example, if f is a simple proc in *Maple*, then the following *Maple* version of df could be used in a similar manner.

```
with(student);
df:=proc(f) local dq;
dq:=(f(x+.00001)-f(x))/.00001;
makeproc(dq,x);
end;
```

Integration is more difficult. The idea of defining a function by using the definite integral with one limit of integration fixed and the other allowed to vary is a major stumbling block for calculus students. In our treatment of integration, students have written a program called Riem which accepts a function and a pair of numbers, and computes an approximation to the integral of the function over the interval determined by the points. The students are then asked to write the following program.

```
Int :=   func(f,a,b);
            return func(x);
                    if a <= x and x <= b then return Riem(f,a,x);
            end;
        end;
```

Using this program, students are able to construct and study approximations to the logarithm function and inverse trigonometric functions.

Schemas. Our use of programming activities to help students form schemas to organize collections of individual constructs and other schemas is, at the time of this writing, somewhat ad hoc. Some progress is being made, and has been reported in [2] for a limit schema. Roughly speaking, we ask students to write a set of computer programs that implements a mathematical concept and then to apply their code to specific situations.

For example, developing a schema for the Fundamental Theorem of Calculus requires very little code but it is extremely complicated. Having written the two computer functions df to approximate the derivative and Int to approximate the integral (see above), students are asked to write code that will first do one and then the other, in both orders. This problem gives the students considerable difficulty and they struggle with it for a long time. We feel that this is a useful struggle because it has to do with their ability to interpret functions as objects, to develop processes corresponding to differentiation and integration, and to put it all together in what is essentially a statement of the Fundamental Theorem.

The actual code to solve this problem is very short:

```
> df(Int(f(a,b)));
```

```
> Int(df(f),a,b);
```

We ask students to apply their code to a specific function and to construct a table with four columns: values of the independent variable, corresponding values of f, and corresponding values of the above two lines of code. When the example is a function that does not vanish at a, then the second and third columns are identical, but the fourth is different. The students see the point right away—all three columns are supposed to be the same, but they feel they have made an error in connection with the last column. After some investigation, many students tend to discover on their own the idea of the "constant of integration."

We should note here that it is not our intention to suggest that the approach we are describing is the *only* way to help students understand the ideas surrounding the Fundamental Theorem of Calculus, such as velocity and accumulation and the relationship between them. There are other approaches, such as that being pursued by Jim Kaput [18] in which children control simulated motions on a computer screen. In this case, the functions are defined graphically rather than analytically. It would be interesting to see whether the theoretical framework we are using would apply in the same way to describe student understanding of functions.

Sometimes we don't ask students to write code but rather to investigate code which we provide. We do this in situations where the particular code involves more in the way of programming issues than mathematical issues. This is the case for the following example which simulates the operation of induction. This code makes use of the computer function, `implfn`, which they have written (see above) and is applied to a boolean-valued function P. The first few lines find a starting point and the rest of the code runs through the induction steps. If the proposition does hold from the selected starting point on, then the code will run forever.

```
start := 1;
while P(start) = false do
    start := start + 1;
end;


L := [];
n := start; L(n) := true;
while L(n) = true and implfn(P)(n) = true do
    L(n+1) := true;
    n := n+1;
    print "The proposition P is true for n = ", n;
end; print "P is not proven for n = ", n+1;
```

Making the abstract concrete. A second way in which working with an appropriate computer language can help students construct mathematical concepts is that the computer can provide an environment where students are able to make certain abstract notions concrete. Consider, for example, the statement that a function f maps its domain D onto a set S:

For each $y \in S$ there exists an x in D such that $f(x) = y$.

Students may consider such a precise definition to be a difficult abstraction. They can be helped by working with a *MPL* such as *ISETL* in which such a statement can be made, run, tested, and reasoned about. Following is *ISETL* code that expresses the same mathematical statement.

```
forall y in S | (exist x in D | f(x)=y);
```

If S, D, and f are defined, then this code can be run to return the value **true** or **false**.

Our way of using this feature in instruction is to begin with a somewhat vague discussion of the essential idea, perhaps in the context of students working on a problem that requires this idea. Then we ask students to write a program such as the above and use this program in solving problems. It appears that writing and working with a program that he or she has written helps the student make concrete the ideas embodied in the program.

Indirect effects of working with the computer. One example of an indirect effect on students is reported in [1]. Students were asked at the beginning of a discrete mathematics course to give examples of functions. A very high percentage of their responses were functions defined by simple expressions such as $x^2 + 1$ and many students displayed no more than an action conception of function, or even no useful conception at all. After several weeks of work with a mathematical programming language involving procedures, sets, and finite sequences, but before any explicit study of functions in the course, students were asked again to provide examples. This time the examples given were much richer and more varied, with a number of students' responses indicating they were moving from action conception or no conception to a process conception of function.

3.3 Data.

The third component of the framework has to do with the collection and analysis of data.

There are several kinds of data which must be gathered in studies under this framework. There must be information about the students and the course(s) taken. In some cases, we gather data about students who have previously studied the mathematics we are concerned with under traditional instruction. In other cases, we will study students who have experienced the kind of instruction we describe in this paper. Where appropriate, we will summarize rough comparisons of performances of the two kinds of students. Sometimes, students' attitudes about mathematics and about the particular subject matter is of concern in the study. Most important, of course, is data that allows us to analyze the students' relationship to the particular material; it is this last category that we will consider now in some detail.

3.3.1 Forms of data.

We see two reasons to gather a number of different kinds of data about the students' relationship to the material. First, the methods used do not provide precise information leading to inescapable conclusions. The best we can hope for is data that is illustrative and suggestive. Our confidence in any tentative conclusions that might be indicated is increased as we widen the sources of information about student knowledge and how it develops. The second reason is that we are interested in two different kinds of issues: what mental constructions students might be making and how much of what mathematics do they seem to be learning and using? That is, we are comparing the mental constructions the students appear to be making with those called for in the theoretical analyses, and at the same time we are searching for the limits of student knowledge. Often, the same information will shed light on both questions, but sometimes it is necessary to use different kinds of data to investigate these different issues.

In our studies we gather data using three kinds of instruments: written questions and answers in the form of examinations in the course or specially designed question sets; in-depth interviews of students about the mathematical questions of concern; and a combination of written instruments and interviews. For purposes of data analysis, all of our data is aggregated across the set of students who participated in the study.

The written instruments contain fairly standard questions about the mathematical content and they are analyzed in relatively traditional ways. We grade the responses on appropriate scales from incorrect to correct with partial credit in between, and then count the scores. Where appropriate we list the specific points (both correct and incorrect) in the responses of all the students and collate those points. This information tells us about what the students may or may not be learning and also about possible mental constructions.

The interviews of students form the most important and the most difficult part of our observation and assessment activity. The audio-taped transcripts of the interviews complement the record of written work which the student completes during the interview. An even more complete picture might be obtained by videotaping the interviews, but we have not yet added this component to our interviews. One reason why the interviews are far more valuable than written assessment instruments used alone is that for one student the written work may appear essentially correct while the transcript reveals little understanding, while for another student the reverse may be true. There is no set recipe for designing interview questions. The research team proposes, discusses, and pilots questions intended to test the hypotheses set forth by the current version of the genetic decomposition for the topic under study. An instrument is then put together which is administered to a number of students by a number of researchers. Communication among the interviewers before and between interviews is important to increase consistency.

A second combination of written instruments and interviews is used in the following way. The written instrument is administered to a total population and the responses are used in designing interview questions. For example, the student might be asked in the interview to explain what was written, or if he or she wished to revise the response. If the written instrument were administered to a group (as in a group examination such as described in [23]), we would ask whether the interviewee was fully in accord with and understood the group response. Individuals are selected for interview based on their responses. The idea is that it may not be necessary to interview all of the students who gave a certain written response. In selecting students to interview, we try to access the full range of understanding by including students who gave correct, partially correct, and incorrect answers on the written instruments. We also routinely select students who appear to be in the process of learning some particular idea rather than those who have clearly mastered it or those who had obviously missed the point. In this way, it is possible to interview only a small percentage of the total population, but still investigate every written response that appeared. Usually it is feasible, for each specific response, to interview more than one student who made that response. A combination of the two kinds of instruments is discussed in [10].

3.3.2 Analyses of interview data.

One of the most serious practical difficulties in doing qualitative research is the very large amount of data that is generated and that must be analyzed. We believe that our framework offers an alternative to the approach used by some researchers (see, for example, [27]) who attempt to make a full analysis of the total set of data. In the following paragraphs we begin by discussing the goals of our analyses of interviews and then describe the specific steps by which we attempt to achieve these goals.

Goals of the data analyses Consider in Figure 1 the bi-directional arrow connecting the theoretical analyses with the observations and assessments. What the data tells us can support or lead to revisions of the particular analyses that have been made of the concept being studied, and even the general theoretical perspective. This is the meaning of the arrowhead pointing up towards the theoretical analysis in the figure.

On the other hand, the arrowhead in the opposite direction, pointing down, indicates our method of using the analysis of the concept to focus our investigation of the data. In other words, the theoretical analysis tells us what questions to ask of the data. More specifically, our study of the data is narrowed by focusing on the question of whether the specific constructions proposed by the theoretical analysis (which are the main determinant of the design of instruction) are in fact being made by the students who succeed in learning the concept. Put another way, we ask if making, or failing to make, the proposed constructions is a reasonable explanation of why some students seemed to learn the concept and others did not.

Of course, actual student learning is seldom characterizable in binary, yes/no terms. Students range on a spectrum from those who seem to understand nothing (about the particular piece of mathematics) to those who indicate a mature understanding compatible with the understanding of mathematicians. The goal of our analysis of the data is to establish a parallel spectrum of mental constructions, going from those who appeared to construct very little, through those who constructed bits and pieces, to those who seemed to have made all of the constructions proposed by the theoretical analysis.

It is easy to see how such an approach requires iterating through the steps in Figure 1 as the parallel spectra of mathematical understanding and mental constructions are unlikely to be completely similar initially and the researcher must endeavor, in the repetitions, to try to bring them in line with each other.

Steps of the data analyses. Our interviews are audio-taped and transcribed. These transcriptions, together with any writing performed by the interviewee and any notes taken by the interviewer, make up the data which are to be analyzed. We do this in five steps.

1. **Script the transcript.** The transcript is put in a two-column format. The first column contains the original transcript and the second column contains an occasional brief statement indicating what is happening from that point until the next brief statement. It is convenient to number the paragraphs at this point.

2. **Make the table of contents.** A table of contents is constructed. The statements in the scripting should be a refinement of the items in the table of contents.

These first two steps are designed to make the transcript more convenient to work with and to give the researcher an opportunity to become familiar with its contents.

3. **List the issues.** By an issue we mean some very specific mathematical point, an idea, a procedure, or a fact, for which the interviewee may or may not construct an understanding. For example, in the context of group theory one issue might be whether the student understands that a group is more than just a set, that is, it is a set together with a binary operation.

The researcher begins to generate the list of issues by reading carefully through the transcript for each interviewee writing down each issue that seems to be discussed and noting the page numbers (or paragraph numbers) where it appears. These lists of issues for individuals are then transposed to form a single list of issues and, for each, a list of the specific transcripts (and location) in which it occurred. We believe that the best results are obtained at this step if these lists are generated independently by several researchers with subsequent negotiations to reconcile differences. Since the list of issues varies widely from one set of interviews to another, and since the issues are often not characterizable in terms of number or type of occurrence, we have not attempted to produce information such as inter-rater reliabilities. However, we have found that in most cases, the various researchers independently produce lists of issues which are very similar, and that differences in these lists are often a matter of a single issue being referred to by several different names. Negotiation amongst the researchers serves both to reconcile these differences and to clarify the issues and the terminology being used to describe them.

At this point a selection is made. If an issue occurs for only a very small number of interviewees, and at only a few isolated places, then it is unlikely that this data will shed much light on that issue. There are several reasons why this might occur. Perhaps no student came close to understanding the concept; perhaps all students were well beyond their struggles to understand that concept; or, it is possible that the interview questions were not successful in getting many students to confront these issues very often. In the latter case, it is natural to relegate the issue to future study.

The research team chooses for further study those issues which occur the most often for the most interviewees, and for which there is the largest range of successful, unsuccessful, and in-between performance to be explained.

4. **Relate to the theoretical perspective.** Each issue is considered in detail. The researchers try to explain the differences between the performances of individual students on the issue in terms of whether they constructed (or failed to construct) the actions, processes, objects, and schemas proposed by the theoretical analysis. If it is necessary to bring the theoretical analysis more in line with the data, the researchers may drop some constructions from the proposed genetic decomposition, or look for new ones to add to the theoretical analysis.

Focusing on the successes which have occurred, the researchers attempt to reconcile the way in which successful students appear to be making use in their thinking of the constructions predicted by the theoretical analysis. Again, adjustments in that analysis are made as necessary.

If the data appear to be too much at variance with the general theoretical perspective, consideration is given to revisions of the perspective. This can take the form of adding new kinds of constructions, or revising the explanations of constructions already a part of the theory.

If drastic changes are required very often, or if each new iteration of the framework continues to require major changes and the process is more like an oscillation than a convergence, then consideration must be given to rejecting the general theoretical perspective.

In each of these steps, each member of the research team makes an independent determination and all differences are reconciled through negotiation. Nothing appears in the final report that is not agreed to by all authors. This is the closest we come to objectivity in these considerations, and it is one reason why our papers tend to have long lists of authors.

5. **Summarize performance.** Finally, the mathematical performance of the students as indicated in the transcripts is summarized and incorporated in the consideration of performance resulting from the other kinds of data that were gathered.

4. Discussion of accuracy and assessment

We return now to the issues raised in Section 2.2.

We have already discussed the development of our theoretical perspective beginning with its origins in the ideas of Piaget and as a result of our work within the framework (Section 2.1). We also explained in some detail how the theoretical analysis influences our instructional treatment (Section 3.2) and the relationship between the theoretical analysis and the analysis of data (Section 3.3.2). Regarding the use of data from other studies, our approach is to use all data that is available to the researchers at the time that they make their analyses.

There are two main issues that remain to be discussed: the relation between our theoretical perspective and accuracy about what is going on in the mind of the student, and the question of assessment, including a consideration of the circumstances under which our theory could be falsified.

4.1 Our theory and reality.

It is important to emphasize that, although our theoretical analysis of a mathematical concept results in models of the mental constructions that an individual might make in order to understand the concept, we are in no way suggesting that this analysis is an accurate description of the constructions that are actually made. We believe that it is impossible for one individual to really know what is going on in the mind of another individual. In this respect our theoretical framework is like its underlying radical constructivist perspective which von Glasersfeld notes, "is intended, not as a metaphysical conjecture, but as a conceptual tool whose value can be gauged only by using it" [14]. All we can do is try to make sense out of the individual's reactions to various phenomena.

One approach would be to try to make inferences from these reactions about the actual thinking processes of the respondent. We reject this because there is no way that we could check our inferences. Rather, we take something of the view of von Glasersfeld [13] and consider only whether our description of the mental constructions is compatible with the responses that we observe. That is, we ask only whether our theoretical analysis is a reasonable explanation of the comments and written work of the student. With respect to the instructional treatment, we confine ourselves to asking whether those strategies that are derived from our explanations appear to lead to the student learning the mathematical concept in question.

4.2 Assessment.

We make both an internal and external assessment of our results as the work proceeds.

Internally, we ask whether the theoretical analysis and resulting instruction "work" in the sense that students do (or do not) appear to be making the mental constructions proposed by the theory. That is, are the students' responses reasonably consistent with the assertion that those mental constructions are being made?

Externally, we ask if the students appear to be learning the mathematical concept(s) in question. We ask and answer this question in more or less traditional terms through the results of examinations and performance on the mathematical questions in our interviews.

Finally, at the extreme end of assessment is the question of falsification. Any scientific theory must contain within it the possibility that an analysis, or even the entire theory should be rejected. In our framework, revisions including major changes in, or even rejection of, a particular genetic decomposition can result from the process of repeating the theoretical analyses based on continually renewed sets of data. As we indicated above (Section 4), the need for continual and extensive revisions could lead to complete rejection of the general theory and, presumably, the entire framework with which we are working.

In a more positive vein, the continual revisions of our theoretical analyses and the applications to instruction in ongoing classes with, presumably, successful results over a period of time tends to ensure that the longer the work continues, the less likely that our framework and the general theoretical perspective it includes will turn out to have been totally useless. This is not to say, of course, that in the future, some more effective and more convenient framework and/or theory might not emerge and replace what we are using.

This paper has focused on the research methodology in a general approach to research and development in undergraduate mathematics education. It is closely related and complementary to two other papers about this same approach: [8] and [9]. The former concentrates on a somewhat historical description of the programming language *ISETL* and the latter focusses on the pedagogical aspect.

5. Conclusion

In conclusion, we would like to make one final point about the use of the framework described in this paper. Throughout our discussion we have given examples from previous studies in which developing forms of this framework were applied. These investigations had to do with the concept of function, mathematical induction, and predicate calculus. In subsequent publications, we expect to report on studies of students learning concepts in calculus and abstract algebra. Taken as a totality, these papers express a particular approach to investigating how mathematics can be learned and applying the results of those investigations to help real students in real classes. It is our intention to provide enough information in these reports to allow the reader to decide on the effectiveness of our method for understanding the learning process and for helping students learn college level mathematics.

REFERENCES

1. Breidenbach, D., Dubinsky, E., Hawks, J., and Nichols, D., *Development of the process conception of function*, Educational Studies in Mathematics **23** (1992), 247–285.
2. Cottrill, J., Dubinsky, E., Nichols, D., Schwingendorf, K., Thomas, K., and Vidakovic, D., *Understanding the limit concept: Beginning with a coordinated process schema*, Journal of Mathematical Behavior (in press).
3. Dautermann, J., *Using ISETL 3.0: A Language for Learning Mathematics*, West, St Paul, 1992.
4. Davis, R., *Learning Mathematics: The Cognitive Science Approach to Mathematics Education*, Ablex Publishing Corporation, Norwood, NJ, 1984.
5. Dubinsky, E., *On learning quantification*, Journal of Computers in Mathematics and Science Teaching (in press).
6. Dubinsky, E., *Reflective abstraction in advanced mathematical thinking*, Advanced Mathematical Thinking (D. Tall, ed.), Kluwer, Dordrecht, The Netherlands, 1991, pp. 231–250.

7. Dubinsky, E., *A theory and practice of learning college mathematics*, Mathematical Thinking and Problem Solving (A. Schoenfeld, ed.), Erlbaum, Hillsdale, NJ, 1994, pp. 221–243.

8. Dubinsky, E., *ISETL: A programming language for learning mathematics*, Communications in Pure and Applied Mathematics **48** (1995), 1–25.

9. Dubinsky, E., *Programming to Learn Advanced Mathematical Topics: One of Many Computational Environments* (in preparation).

10. Dubinsky, E., Dautermann, J., Leron, U., and Zazkis, R., *On learning fundamental concepts of group theory*, Educational Studies in Mathematics **27** (1994), 267–305.

11. Dubinsky, E., Schwingendorf, K. E., and Mathews, D. M., *Calculus, Concepts and Computers (2nd ed.)*, McGraw-Hill, New York, 1995.

12. Fisher, Sir R. A., *Statistical Methods for Research Workers*, Oliver & Boyd, Edinburgh, 1932.

13. von Glasersfeld, E., *Learning as a constructive activity*, Problems of Representation in the Teaching and Learning of Mathematics (C. Janvier, ed.), Erlbaum, Hillsdale, NJ, 1987, pp. 41–69.

14. von Glasersfeld, E., *Radical Constructivism: A Way of Knowing and Learning*, Falmer Press, New York, 1995.

15. Griese, D., *The Science of Programming*, Springer-Verlag, New York, 1981.

16. Gutting, G. (Ed.), *Paradigms and Revolutions: Appraisals and Applications of Thomas Kuhn's Philosophy of Science*, University of Notre Dame Press, Notre Dame, IN, 1980.

17. Jacob, E., *Clarifying qualitative research: A focus on traditions*, Educational Researcher **17(1)** (1988), 16–24.

18. Kaput, J. and Roschelle J., *Year 2 Annual Report to NSF: SimCalc Annual Report to NSF: Materials, Tests, and Results*, unpublished report available from first author at Department of Mathematics, UMass Dartmouth, No. Dartmouth, MA 02747, 1995.

19. Kuhn, T., *The Structure of Scientific Revolutions*, (2nd ed.), University of Chicago Press, Chicago, 1970.

20. Patton, M. Q., *Qualitative Evaluation and Research Methods*, Sage Publications, Newbury Park, 1990.

21. Piaget, J., *The Principles of Genetic Epistemology*, (W. Mays, Trans., original work published 1970), Routledge and Kegan Paul, London, 1972.

22. Piaget, J. and Garcia, R., *Psychogenesis and the History of Science*, (H. Feider, Trans., original work published 1983), Columbia University Press, New York, 1989.

23. Reynolds, B., Hagelgans, N., Schwingendorf, K., Vidakovic, D., Dubinsky, E., Shahin, M., and Wimbish, G., *A Practical Guide to Cooperative Learning in Collegiate Mathematics*, (MAA Notes Number 37), The Mathematical Association of America, Washington, DC, 1995.

24. Romberg, T., *Perspectives on scholarship and research methods*, Handbook of Research on Mathematics Teaching and Learning (D. A. Grouws, ed.), Macmillan, New York, 1992, pp. 49–64.

25. Schoenfeld, A. H., *On having and using geometric knowledge*, Conceptual and procedural knowledge: The case of mathematics (H. Hiebert, ed.), Erlbaum, Hillsdale, NJ, 1986, pp. 225–264.

26. Schoenfeld, A. H., *Some notes on the enterprise (Research in collegiate mathematics education, that is)*, CBMS Issues in Mathematics Education: Research in Collegiate Mathematics Education. 1 **4** (1994), 1–19.

27. Schoenfeld, A. H., Smith, J. P., III, and Arcavi, A., *Learning: the microgenetic analysis of one student's understanding of a complex subject matter domain*, Advances in Instructional Psychology, 4 (R. Glaser, ed.), Erlbaum, Hillsdale, NJ, 1993.

28. Sfard, A., *Two conceptions of mathematical notions, operational and structural*, Proceedings of the Eleventh Annual Conference of the International Group for the Psychology of Mathematics Education (A. Borbàs, ed.), University of Montreal, Montreal, 1987, pp. 162–169.

29. Sfard, A., *On the dual nature of mathematical conceptions*, Educational Studies in Mathematics **22** (1991), 1–36.

30. Suppe, F. (Editor), *The Structure of Scientific Theories*, 2nd Ed., University of Illinois Press, Urbana, IL, 1977.

31. Tall, D. and Vinner, S., *Concept image and concept definition in mathematics with particular reference to limits and continuity*, Educational Studies in Mathematics **12** (1981), 151–169.

32. Vidakovic, D., *Differences between group and individual processes of construction of the concept of inverse function*, (unpublished doctoral dissertation), Purdue University, West Lafayette, IN, 1993.

Acknowledgments. The authors would like to express their most sincere gratitude to two groups of people. First, to all of the researchers whose works we have cited in this monograph for contributing immeasurably to our collective understanding. Second, to the members of our Research in Undergraduate Mathematics Education Community (RUMEC) in general with special thanks to Bernadette Baker, Julie Clark, Jim Cottrill, and Georgia Tolias for their thoughtful comments and criticisms of earlier drafts of this manuscript. Finally, we are grateful to James Kaput and Alan Schoenfeld for their numerous and penetrating suggestions.

This work was partially supported by grants from the National Science Foundation, Division of Undergraduate Education (DUE) and the Exxon Education Foundation.

PURDUE UNIVERSITY, WEST LAFAYETTE, INDIANA

INDIANA UNIVERSITY, SOUTH BEND, INDIANA

GEORGIA COLLEGE, MILLEDGEVILLE, GEORGIA

PURDUE UNIVERSITY, WEST LAFAYETTE, INDIANA

CENTRAL MICHIGAN UNIVERSITY, MOUNT PLEASANT, MICHIGAN

UNIVERSITY OF WISCONSIN–PLATTEVILLE, PLATTEVILLE, WISCONSIN

CBMS Issues in Mathematics Education
Volume **6**, 1996

The Creation of Continuous Exponents: A Study of the Methods and Epistemology of John Wallis

DAVID DENNIS AND JERE CONFREY

I learned empirically that this came out this time, that it usually does come out; but does the proposition of mathematics say that? I learned empirically that this is the road I traveled. But is *that* the mathematical statement?—What does it say, though? What relation has it to these empirical propositions? The mathematical proposition has the dignity of a rule.

Ludwig Wittgenstein [29, p. 47e]

Introduction

Within a constructivist framework, we study student conceptions in order to develop approaches to concepts that make use of students' inventions and their creative use of resources. Our goal is to improve their understanding of mathematical ideas. However, we find that researchers who are traditionally educated in mathematics often fail to recognize or legitimate student methods, and need to broaden their understanding of possible routes towards the development of an idea. (We include ourselves in this description.) History has proven ideal for this, because it provides rich sources of alternative conceptualization and diverse routes to the development of an idea. Thus, we have found it to be a provocative and stimulating source of preparation for "close listening" to student mathematics [10].

Moreover, such historical research invariably goes well beyond this original description. Our historical research guides us in building provocative tasks for

This paper was originally presented at the symposium *The Function Concept and the Development of a Constructivist Research Program* at the annual meeting of the American Educational Research Association, April 12–16, 1993 in Atlanta, GA. This research was funded under a grant from the National Science Foundation (MDR 9245277).

our interviews. It leads us to make alternative proposals for curricular and instructional development. And it provides opportunities for teacher education: through the exploration of historical example, we can assist teachers in gaining depth in and perspective about their content knowledge. But, perhaps most of all, our historical work leads us to reconceptualize our beliefs about the epistemology of mathematics. Our historical work effects our epistemological perspective, and our epistemological perspective influences the way that we engage in and interpret history. We would argue that the circularity in this argument is not a weakness but a necessity. Historical work serves to inform us about the present—many of our current assumptions come to light through historical work. We recognize that there is no way that historical work can profess ultimate accuracy when recast by modern scholars. However, by use of original texts (when possible) and locating the work within the socio-cultural and historical context and by assuming a pluralistic history, we can attempt to understand it from the perspective of the originators. The methodological standards which are applied, are derived with respect to a set of epistemological assumptions which guide our overall research program.

The broad epistemological assumptions which underlie this historical investigation include the following:

1. A Lakatosian viewpoint that the development of mathematical ideas is characterized as a series of conjectures and refutations. Lakatos documented the limitations of a formalist methodology that hides the bold conjectures and masks the debates and refutations to present only the final product of a work. We take the view that mathematics is the process of proposing, developing, modifying and revising one's ideas, and that the proof, a normative assessment signaling the community's acceptance of an idea represents only a part of that process. Near the end of his life, Lakatos began to describe mathematics as quasi-empirical by which he referred to its progress via a set of heuristics. Whereas we understand and agree with Lakatos's concern that mathematics be viewed as more than a set of Euclidean axiomatics, we find his use of the term "quasi-empirical" somewhat ironic.

Much mathematics developed as a means to describe what scientists did and as a result, it is unclear where to draw the line between quasi-empirical and empirical. The modern mathematician R. Courant (1888–1972) wrote of C. F. Gauss that:

> "He (Gauss) was never aware of any contrast, not even of a slight line of demarcation, between pure theory and applications. His mind wandered from practical applications, undaunted by required compromise, to purest theoretical abstraction and back, inspiring and inspired at both ends. In light of Gauss' example, the chasm which was to open in a later period between pure and applied mathematics appears as a symbol of limited human capacity. For us today as we suffocate in specialization, the phenomenon of

Gauss serves as an exhortation . . . it is critical for the future of our science that mathematicians adopt this course, both in research and in education." [13, p. 132]

2.   As an extension to a Lakatosian viewpoint, we view the history of mathematics as the coordination and contrast of multiple forms of representation. In history often one sees a particular form of representation as primary for the exploration, whereas another might form the basis of comparison for deciding if the outcome is correct. The confirming representation should be relatively independent from or contrasting to the primary exploratory representation. It must show contrast as well as coordination for the insight to be compelling. For example, Descartes most often generated his curves as loci of points tracing out certain movements in the plane. Algebra was used as the means for describing the curve, and as a result, he built his axis to be convenient for this description. The axes could be perpendicular or skewed depending on the situation. If a similarity argument was devised, a skewed axis was often built parallel to some line in the construction; if the Pythagorean theorem was used, a perpendicular axis was constructed. (See Smith, Dennis and Confrey [25] or Dennis [14] for a further discussion of this.) Thus, the modern Cartesian plane was not the starting point for Descartes' plane but was generated as a descriptive alternative representation. This sense is lost and distorted in our current curriculum where the Cartesian plane is treated predominantly as a means of displaying algebraic equations. This misrepresentation of history is not atypical. In contrast, we propose to use the idea of an *epistemology of multiple representations* (Confrey [9] and Confrey and Smith [12], in press). This approach leads us to examine the historical records in order to document such questions as what forms of representation were most influential for a given mathematician, what frameworks were used for confirmation, what representations were available as a part of "standard knowledge" for the time period, and how the mathematician moved among these representations to create and modify and extend mathematical pursuits. In reading this paper, one will see how the use of geometric insights into area and volume inform reasoning involving algebraic formula and how these are useful in confirming bold conjectures concerning patterns in a table. These table patterns later create an important transition that opens the door to admit infinite series into mathematical reasoning. Weaving these forms of representation together creates the framework that allows for progressive mathematical thought.

3.   In contrast to what is explicitly developed by Lakatos, we believe that the socio-cultural period of the work exerts significant influence on its mathematical development. When possible within a scanty historical record, we will introduce such information.

Creating a reconstruction of the historical development of mathematical concepts is a difficult task for a number of reasons. These include a relative lack of access to original texts, the difficulty of interpreting original text as it appears notationally, the tendency to find secondary sources that distort the historical record by indiscriminately converting them to modern notation or form (see Unguru [27] for such a criticism) and a relative paucity of historical work in mathematics.[1]

We set our task in this paper to describe the development of rational exponents. We wished to understand this development at a deeper level than to simply assume that the development of rational exponents was a matter of extending a number pattern and its properties. What we found was that history reveals differences in the use of the ideas of powers, indexes, and exponents. In order for a generalized concept of continuous exponents to become accepted as legitimate mathematics it first had to be validated across several representations: table arithmetic, geometry, and algebra. The story of the creation of continuous exponents is linked strongly to the evolution of the notions of areas, limits of ratios, ratios with negative numbers, and continuous functions in general. Without this wider context only positive integer powers were accepted by a generation of mathematicians.

4. We did this historical research in light of the "splitting conjecture" [8, 11]. According to this conjecture, Confrey postulated a different cognitive basis for splitting vs. counting. She suggested that basing multiplication in schools on repeated addition was neglecting the development of a parallel but related idea of equal sharing, of reproduction, magnification, etc. Her splitting world includes the interrelated development of ratio, similarity, multiplication, division, multiplicative units, rates and exponential functions. Using this perspective, we conducted our historical investigation considering carefully how the use of geometry and ratio enlightened the development of mathematical thinking, seeking to avoid masking those distinctions in a generic algebraic description.

Historical Background

The following is a work of historical investigation that focuses on specific mathematical moments where methods, concepts, and definitions underwent profound changes. There are two main goals. First to sketch the history of the development of a continuous concept of exponents, and second to examine carefully the epistemological setting in which these developments took place. Put more simply, what was it that convinced certain mathematicians that their concepts were viable. It is not intended to be a complete historical discussion, but rather a series of illuminating snapshots. The mathematical details of these moments are provided to the extent necessary for an understanding of the epistemology.

The main focus of this paper is on the mathematical methods of John Wallis (1606–1703), and in particular his influential work, the *Arithmetica Infinitorum* [28], first published in 1655. Although he was not the first person to suggest the

---

[1]For a good example of the kind of historical work that is required see Fowler [18].

use of fractional exponents, his work provided compelling reasons for their adoption. After reading Wallis, the young Isaac Newton (1642–1722) was inspired to derive his general binomial series [30]. The binomial series was, in turn, the main tool used by Leonard Euler (1707–1783) to explore the world of continuous functions including natural based exponentials and logarithms [17]. The work of all three of these men was carried out in an empirical setting without recourse to formal logical proofs. They checked and double checked different representations against each other until they were convinced of the validity of their results. The formal proofs of the nineteenth century grew out of these results but often mask the methods.

Wallis openly advocated an empirical or heuristic approach to mathematical truth. He became convinced of the validity of his mathematics through a series of conjectures and confirmations. His main arguments depended upon a coordination of multiple representations. These included numerical tables, algebra, and geometry. For Wallis, a definition became reasonable when it emerged as a pattern in one representation but could also be confirmed through agreement with another. Just that an idea was reasonable in one setting was never enough for Wallis. His primary investigations often took place in the setting of numerical sequences and tables. He then sought confirmation through algebra and geometry.

Wallis practiced induction, and by this word he did not mean formal mathematical induction, but informal or scientific induction. That is to say, he sought a pattern, checked a series of examples, and then assumed his rule was valid so long as he found "no ground of suspicion why it should fail" [23, p. 385]. Formal proofs by mathematical induction were being carried out by his contemporaries Fermat (1601–1665), and Pascal (1623–1662). Fermat criticized the methods of Wallis as suggestive but incomplete. Wallis responded, saying that he was trying to develop a theory of knowledge that was far superior to the logical analysis of known results. He claimed that Fermat "doth wholly mistake that treatise [i.e. the *Arithmetica Infinitorum*] which was not so much to show a method of Demonstrating things already known as to show a way of Investigation or finding of things yet unknown" (quoted in [23, p. 385]). Wallis felt that the ancient mathematicians were in possession of such a method but that it was "studiously concealed" [23, p. 385] and covered over with logical analysis.

Wallis was concerned with action rather than logical justification. In this he was part of a general philosophical movement away from the Greek-based, neoplatonic thought of the Renaissance. In the seventeenth century Roman texts became far more popular than Greek ones. The works of Seneca and other Roman stoic philosophers were revived. Roman thought is generally much more practical and empirical than Greek thought. Clear and direct language is highly valued.

Wallis was an expert in many languages. Among others he mastered Latin, Greek, Hebrew, and probably Arabic. He first came to prominence during the civil wars in England. He aided the Parliamentary Party by rapidly deciphering coded Royalist messages that had been captured in 1642 [24, p. 6]. This service got him a post at Oxford during the rule of Oliver Cromwell. An instinct for patterns and language marks all of his work. As you read the work of Wallis consider how a cryptographer works. He need not provide logical proof that his interpretation is correct, rather, when an interpretation makes sense his work is finished.

In order to understand the portions of Wallis that we will present it is necessary to know the polynomial formulas for the summation of the integer powers of the first n integers. These formulas just appear in Wallis's *Arithmetica Infinitorum* [28] with no mention of where he got them or how to derive them. Mathematicians in the seventeenth century rarely provided any references, and historians have speculated on Wallis's various possible sources for these formulas. They appear in a number of seventeenth century French sources including works by Fermat and Pascal [4]. Another interesting possible source for Wallis may have been his reading of Arabic texts. Wallis did read Arabic, and these formulas appeared in the work of Abu Ali al-Hasan ibn al-Hasan ibn al-Haitham (965–1039), known in the West as Alhazen. Wallis definitely had access to some of the mathematical work of Alhazen, but historians have not been able to directly verify exactly which texts he read.[2]

Since the focus of this paper is to use historical material for educational purposes, we will begin with a derivation of these summation formulas that is based on a text by Alhazen (Baron [1], Edwards [16]). This method for deriving these formulas fits beautifully with the concepts that appear in the *Arithmetica Infinitorum*, although only circumstantial evidence exists as to what sources Wallis actually read.[3] When discussing the work of Wallis, we will stay very close to the form and notation of the original works. The work of Alhazen, however, will not be discussed in its original form. The Arab mathematicians did not develop symbolic algebra, thus all of the formulas of Alhazen would have been written out in words. For brevity, the derivations of Alhazen are given in modern algebraic form.

It must be noted here that even our presentation of the derivations of Alhazen is somewhat controversial. The geometric figures that we will present in the next section do not actually appear in the Arabic texts of Alhazen. Some historians (Katz [19]) have interpreted this portion of a work by Alhazen as purely arithmetical, while others (Baron [1]) have interpreted this work geometrically, as we will do. In an age of handwritten manuscripts, many geometrical works (including Euclid) often lacked figures, containing instead instructions for the construction of figures. One simply could not trust scribes to accurately reproduce figures. These summation formulas are derived in the beginning of

---

[2]Recently Victor Katz has investigated this question in detail.

[3]Victor Katz, for example, is inclined to believe that Wallis got these formulas from the work of Faulhaber, but then the question arises as to what sources were used by Faulhaber.

a work by Alhazen that is purely geometrical (concerned with the volumes of paraboloids), and in general the work of Alhazen is much more geometrical than other Arabic mathematicians of his time. For these reasons, it seems to us that Baron's geometrical interpretation of these formulas is reasonable.

Putting aside these questions of historical authenticity, the main reason that we have chosen to begin with these derivations springs from our educational focus. If students are to enter into the table interpolations of John Wallis and really understand their significance, they must be able to understand and accept the summation formulas that he took for granted. Of all of the ways that we know of to derive these formulas, this opening interpretation of a work of Alhazen seems best suited to our educational purposes, since it produces these algebraic formulas by linking them to a geometrical representation.

Alhazen's Summation Formulas and Powers Higher than Three

Alhazen was a physicist as well as a mathematician. His work was grounded in the Greek geometrical tradition, but he also sought after practical empirical results in optics and astronomy. Greek mathematics (with the exception of Diophantus and Heron) does not mention any powers higher than three because they could not be directly interpreted in geometry, i.e. as lengths, areas, or volumes. Alhazen wanted to calculate new results concerning areas and volumes which involved the summation of powers higher than three. For example he computed the volume generated by rotating a parabola about a line perpendicular to its axis of symmetry (see Figure 1). For a modern student this would involve integrating a fourth power polynomial. Alhazen stated that the volume is 8/15 of the volume of the circumscribed cylinder. In the tradition of Eudoxus, he set up upper and lower sums of cylindrical slices, and then let the slices get finer and finer. Since the radius of each cylindrical slice follows a square function, the areas follow a fourth power. The sum of the areas of these slices involves summing fourth powers (Edwards [16, p. 85].

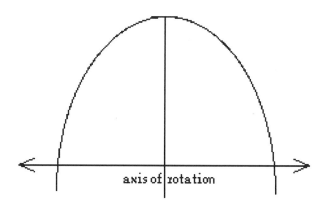

axis of rotation

FIGURE 1

Alhazen derived formulas for the sums of higher powers by coordinating a geometrical interpretation with a numerical representation. His use of areas to represent third and fourth powers broke with the strict geometrical interpretations in Greek mathematics. His method can be used to find a formula for the sum of the powers of the first n integers for any positive integer power. His extension of the concept of powers made sense arithmetically, but was validated through his derivations of new geometric results.

Alhazen derives his formulas by first laying out a sequence of rectangles whose areas represent the terms of the sum (Edwards [16, p. 84]. A rectangle of area a^k is formed using sides of length a^{k-1} and a. He then fills in the rectangle with a series of interlocking strips (see Figures 2, 3, and 4, which are to scale). Alhazen then sets the product of the dimensions of the rectangle equal to the sum of its rectangular parts. Each of the formulas can then be derived from the previous ones. The strips on top, however, involve a double summation in all but the first derivation.

To obtain a formula for the sum of the first n integers, see Figure 2.

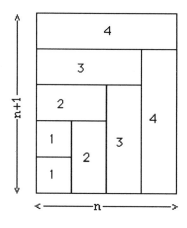

FIGURE 2

$$n(n+1) = \sum_{i=1}^{n} i + \sum_{i=1}^{n} i$$

(1) $$\frac{1}{2}n(n+1) = \frac{1}{2}n^2 + \frac{1}{2}n = \sum_{i=1}^{n} i = 1 + 2 + 3 + \cdots + n$$

To obtain a formula for the sum of squares of the first n integers, see Figure 3.

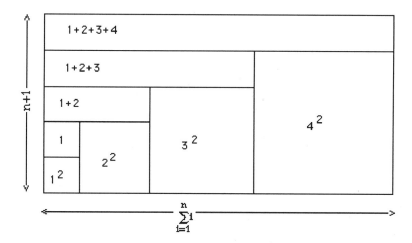

FIGURE 3

$$\left(\sum_{i=1}^{n} i\right)(n+1) = \sum_{i=1}^{n} i^2 + \sum_{i=1}^{n}\left(\sum_{k=1}^{i} k\right)$$

$$\left(\frac{1}{2}n^2 + \frac{1}{2}n\right)(n+1) = \sum_{i=1}^{n} i^2 + \sum_{i=1}^{n}\left(\frac{1}{2}i^2 + \frac{1}{2}i\right)$$

$$\frac{1}{2}n^3 + n^2 + \frac{1}{2}n = \sum_{i=1}^{n} i^2 + \frac{1}{2}\sum_{i=1}^{n} i^2 + \frac{1}{2}\left(\frac{1}{2}n^2 + \frac{1}{2}n\right)$$

$$\frac{1}{2}n^3 + \frac{3}{4}n^2 + \frac{1}{4}n = \frac{3}{2}\sum_{i=1}^{n} i^2$$

(2) $$\frac{1}{3}n^3 + \frac{1}{2}n^2 + \frac{1}{6}n = \sum_{i=1}^{n} i^2 = 1^2 + 2^2 + 3^2 + \cdots + n^2$$

Note that (1) was used twice in obtaining (2).

To obtain a formula for the sum of cubes of the first n integers, see Figure 4.

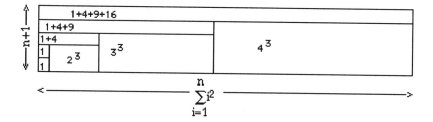

FIGURE 4

$$\left(\sum_{i=1}^{n} i^2\right)(n+1) = \sum_{i=1}^{n} i^3 + \sum_{i=1}^{n}\left(\sum_{k=1}^{i} k^2\right)$$

$$\left(\frac{1}{3}n^3 + \frac{1}{2}n^2 + \frac{1}{6}n\right)(n+1) = \sum_{i=1}^{n} i^3 + \sum_{i=1}^{n}\left(\frac{1}{3}i^3 + \frac{1}{2}i^2 + \frac{1}{6}i\right)$$

$$\frac{1}{3}n^4 + \frac{5}{6}n^3 + \frac{4}{6}n^2 + \frac{1}{6}n = \sum_{i=1}^{n} i^3 + \frac{1}{3}\sum_{i=1}^{n} i^3$$

$$+ \frac{1}{2}\left(\frac{1}{3}n^3 + \frac{1}{2}n^2 + \frac{1}{6}n\right) + \frac{1}{6}\left(\frac{1}{2}n^2 + \frac{1}{2}n\right)$$

$$\frac{1}{3}n^4 + \frac{4}{6}n^3 + \frac{4}{12}n^2 = \frac{4}{3}\sum_{i=1}^{n} i^3$$

(3) $$\quad \frac{1}{4}n^4 + \frac{1}{2}n^3 + \frac{1}{4}n^2 = \sum_{i=1}^{n} i^3 = 1^3 + 2^3 + 3^3 + \cdots + n^3$$

Note that here (2) was used twice and (1) was used once.

A formula for the sum of the first n fourth powers can be derived by continuing this method. First lay out a series of rectangles whose horizontal sides are the cubes and whose vertical sides are once again successive integers. Filling in the strips and proceeding as before will yield the desired formula, although the algebra becomes more tedious. All three of the previous formulas must be used. This derivation is left as an exercise for the reader. This method can be extended to yield a formula for the sum of the first n powers for any integer power. It is a general recursion scheme for these formulas, but at each stage not just one but many of the previous formulas must be used.

For future reference here are Alhazen's Formulas:

(1) $$\qquad 1 + 2 + 3 + \cdots + n = \frac{1}{2}n^2 + \frac{1}{2}n$$

(2) $$\qquad 1^2 + 2^2 + 3^2 + \cdots + n^2 = \frac{1}{3}n^3 + \frac{1}{2}n^2 + \frac{1}{6}n$$

(3) $$\qquad 1^3 + 2^3 + 3^3 + \cdots + n^3 = \frac{1}{4}n^4 + \frac{1}{2}n^3 + \frac{1}{4}n^2$$

(4) $$\qquad 1^4 + 2^4 + 3^4 + \cdots + n^4 = \frac{1}{5}n^5 + \frac{1}{2}n^4 + \frac{1}{3}n^3 - \frac{1}{30}n$$

In order to calculate the volume of the rotated parabola in Figure 1, the actual summation needed is

$$\sum_{i=1}^{n}(n^2 - i^2)^2 = \frac{8}{15}n^5 - \frac{1}{2}n^4 - \frac{1}{30}n$$

which can be derived from (2) and (4) by substituting and collecting terms. The formula inside the summation is the square of the radius of each slice [16, p. 84].

It should be stressed here that results about areas and volumes were not given as formulas but always as ratios. This is true about nearly all such mathematical results until the end of the seventeenth century. For example the area of a triangle is one half of the area of the parallelogram that contains it. The volume of a pyramid is one third of the box that contains it. The area of a piece of a parabola is two thirds of the rectangle that contains it. The area of the circle is $\pi/4$ of the square that contains it. These examples were all known by the second century BC. Alhazen's ratio of $8/15$ is a continuation of that tradition, but he had to coordinate geometry and arithmetic in a new way to find it. In the work of John Wallis we will see an elaborate consideration of area ratios used to justify his definition of fractional and negative exponents. He made extensive use of these summations, and in the final validation of his ideas, Alhazen's ratio of $8/15$ appeared in one of his tables.

The *Arithmetica Infinitorum* of John Wallis

The juxtaposition of arithmetic and geometric sequences goes back at least as far as Aristotle. The idea that one could insert values into such a table by using arithmetic means on one side and geometric means on the other is also ancient. The concept of fractional exponents is suggested in various ways in the fourteenth century works of Oresme and the sixteenth century works of Girard and Stevin (Boyer [2], chapters XIV and XV). The building of tables of logarithms by Napier and others in the early seventeenth century implies the possibility of calculating such powers although this was not how Napier himself interpreted his work. It is important to note that these earlier discussions took place within the world of numerical tables and there was no compelling coordination with geometry. Since geometry was then the dominant form of mathematics, these early works never led to the development or acceptance of fractional exponents.

The Geometry [15], first published in 1638, of René Descartes was the first published treatise to use positive integer exponents written as superscripts. Descartes saw exponents as an index for repeated multiplication. That is to say he wrote x^3 in place of xxx. Wallis adopted this use of an index and tried to extend it, and test its validity across multiple representations.

Wallis took from Fermat the idea of using an equation to generate a curve which was in contrast to Descartes' work which always began with a geometrical construction. Descartes always constructed a curve geometrically first, and then analyzed it by finding its equation (Smith, Dennis, and Confrey [25]). Fermat independently developed coordinate geometry, but his approach was more algebraic (Mahoney [21]). He started with equations and then looked at the curves they generate. This approach is taken by Wallis. His treatment of conic sections, for example, is very similar to that of Fermat (Boyer [3]).

Wallis proposed the notion of fractional indices (exponents) and showed how this notion could be validated in both algebraic and geometric settings [28]. Consideration of areas under curves provided Wallis with a alternative representation with which to validate his proposed arithmetical definition of fractional indices. Wallis's definitions had a lasting impact on mathematics because he demonstrated their viability across multiple representations. Finding the areas under curves was an old problem going back to Archimedes. Beginning in the fourteenth century with Oresme the problem of finding the area under a curve became important because curves were used to represent the magnitudes of velocity over time. The area under the curve then represented the total change in position. This context was laid out in the fourteenth century by Oresme in his *Latitude of Forms* (Calinger [7, p. 224]). In the early seventeenth century Galileo used this concept extensively.

Geometry was considered the primary representation of mathematics in the seventeenth century. This is evident in the work of Barrow and others of that time (Boyer [2]). Arithmetic and algebra were considered, at best, to be forms of shorthand language for the discussion of geometric truth. Some scholars even doubted whether arithmetic and geometry could ever be made consistent. Wallis reverses this order and considers arithmetic as his primary representation (Cajori [6]). In order to validate his arithmetical results he had to show that they yielded the accepted geometric conclusions. To do this he first derived a series of arithmetical ratios and then deduced from them many of the known ratios of areas and volumes.

The *Arithmetica Infinitorum* [28] contains a detailed investigation of the behavior of sequences and ratios of sequences from which a variety of geometric results are then concluded. We shall look at one of the most important examples. Consider the ratio of the sum of a sequence (of a fixed power) to a series of constant terms all equal to the highest value appearing in the sum. Wallis considered ratios of the form:

$$(5) \qquad \frac{0^k + 1^k + 2^k + \cdots + n^k}{n^k + n^k + n^k + \cdots + n^k}$$

For each fixed integer value of k, Wallis investigated the behavior of these ratios as n increases. His investigations are empirical in character. For example when $k = 1$, he calculates:

$$\frac{0+1+2}{2+2+2} = \frac{1}{2} \qquad \frac{0+1+2+3}{3+3+3+3} = \frac{1}{2} \qquad \frac{0+1+2+3+4}{4+4+4+4+4} = \frac{1}{2} \qquad \text{etc.}$$

As n increases this ratio stays fixed at $1/2$. This can be seen from the well known summation formula (1) in a factored form. The numerator is $\frac{n(n+1)}{2}$ while the denominator is $n(n+1)$. Wallis calls $1/2$ the *characteristic ratio* of the index $k = 1$.

When $k = 2$, Wallis computed the following ratios:

$$\frac{0^2 + 1^2}{1^2 + 1^2} = \frac{1}{3} + \frac{1}{6}$$

$$\frac{0^2 + 1^2 + 2^2}{2^2 + 2^2 + 2^2} = \frac{1}{3} + \frac{1}{12}$$

$$\frac{0^2 + 1^2 + 2^2 + 3^2}{3^2 + 3^2 + 3^2 + 3^2} = \frac{1}{3} + \frac{1}{18}$$

$$\frac{0^2 + 1^2 + 2^2 + 3^2 + 4^2}{4^2 + 4^2 + 4^2 + 4^2 + 4^2} = \frac{1}{3} + \frac{1}{24}$$

$$\frac{0^2 + 1^2 + 2^2 + 3^2 + 4^2 + 5^2}{5^2 + 5^2 + 5^2 + 5^2 + 5^2 + 5^2} = \frac{1}{3} + \frac{1}{30}$$

Wallis claimed that the right-hand side is always equal to

$$\frac{1}{3} + \frac{1}{6n}.$$

To see this he applied another summation formula (2) in a factored form. The numerator is always equal to

$$\frac{1}{6}n(n + 1)(2n + 1),$$

while the denominator is equal to $n^2(n+1)$. As n increases this ratio approaches $1/3$, so Wallis then defined the <u>characteristic ratio</u> of the index $k = 2$ as equal to $1/3$. In a similar fashion Wallis computed the characteristic ratios of $k = 3$ as $1/4$, and $k = 4$ as $1/5$. He then made the general claim that the characteristic ratio of the index k is $\frac{1}{k+1}$ for all positive integers k.

In all of his examples Wallis started with the sequence $\{0, 1, 2, 3, 4, \dots\}$ and then raised each term to the index under consideration. Wallis asserted, however, that his values for the characteristic ratios are valid for any arithmetic sequence starting with zero. Changing the difference between the terms would only introduce a constant multiple into all of the terms in both the numerator and the denominator and hence the ratios would remain unchanged. For example, the sum of the squares of any arithmetic sequence (starting at zero) divided by an equal number of terms all equal to the highest term (e.g. $\frac{0^2 + 2^2 + 4^2 + 6^2}{6^2 + 6^2 + 6^2 + 6^2}$) would still approach $1/3$ as more terms are taken.

Wallis then went on to show that these characteristic ratios yielded most of the familiar ratios of area and volume known from geometry. It is here that he showed that his arithmetic was consistent with the accepted truths of geometry. His basic assumptions about the nature of area and volume were taken from Cavalieri's *Geometria Indivisibilibus Continuorum* (1635). He assumed that an area is a sum of an infinite number of parallel line segments,[4] and that a volume is a sum of an infinite number of parallel areas. Wallis first considered the area under the curve $y = x^k$ (see Figure 5). He wanted to compute the ratio of the shaded area to the area of the rectangle which encloses it.

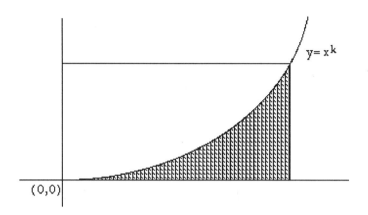

$$y = x^k$$

$$(0,0)$$

FIGURE 5

Wallis claimed that this geometric problem is an example of the characteristic ratio of the sequence with index k. The terms in the numerator are the lengths of the line segments that make up the shaded area while the terms in the denominator are the lengths of the line segments that make up the rectangle (hence constant). Since his ratios are valid for any arithmetic sequence, he imagined the increment or scale as very small while the number of the terms is very large. Hence, for example, the area under a parabola is $1/3$ of the area of the rectangle.[5] This geometric ratio is exactly $1/3$ because the area is made up of an infinite number of line segments. It should be noted here that this characteristic ratio of $1/3$ holds for all parabolas, not just $y = x^2$. For example if we look at the ratio calculation for $y = 5x^2$, both the area under the curve and the area of the rectangle are multiplied by 5, and hence the characteristic ratio of $1/3$ remains the same. Characteristic ratio depends only on the exponent and not on the coefficient. That is to say, *characteristic ratio is not linear.*

This characteristic ratio of x^2 also shows that the volume of a pyramid is $1/3$ of the box that surrounds it (see Figure 6). The pyramid is the sum of a series of

---

[4]This conception of a curve as a series of line segments erected along a line gave rise to the terms "abscissa" and 'ordinate.' Abscissa is Latin for "that which is cut off" and contains the same root as our word "scissors." Ordinates are the ordered series of line segments which are being erected from that which was cut off (abscissa). These terms were coined by Leibniz.

[5]The skeptical reader could fix the interval from 0 to M, and then let $n + 1$ be the number of equal size sub-intervals. Set up the Riemann lower sum divided by the area of the rectangle. Canceling common factors you will arrive at (5). Riemann sums, however, are based on sums of areas of rectangles and this was not Wallis's conception of area.

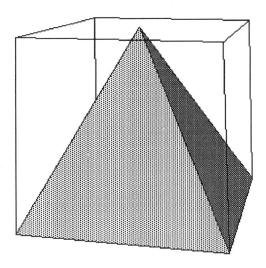

FIGURE 6

squares whose sides are increasing arithmetically. The box is a series of squares whose sides are constant and always equal to the largest square. Hence Wallis sees this as another example of his computation of the characteristic ratio for the index $k = 2$. The same is true for the ratio of the cone to its surrounding cylinder. Here all of the terms both top and bottom are multiplied by π.

These geometric results were not new. Fermat, Roberval, Cavalieri and Pascal had all previously made this claim that when k is a positive integer; the area under the curve $y = x^k$ had a ratio of $\frac{1}{k+1}$ to the rectangle that encloses it (Edwards [16, Chapter 4]. Pascal had given a formal induction proof of this result. Wallis, however, goes on to assert that if we define the index of $\sqrt{x}$ as $1/2$, the claim remains true. Since the area under the curve $y = \sqrt{x}$ is the complement of the area under $y = x^2$ (i.e. the unshaded area in Figure 5), it must have a characteristic ratio of

$$\frac{2}{3} = \frac{1}{\frac{1}{2} + 1}.$$

The same can be seen for $y = \sqrt[3]{x}$, whose characteristic ratio must be

$$\frac{3}{4} = \frac{1}{\frac{1}{3} + 1}.$$

It was this coordination of two separate representations that gave Wallis the confidence to claim that the appropriate index of $y = \sqrt[q]{x^p}$ must be p/q, and that its characteristic ratio must be

$$\frac{1}{\frac{p}{q} + 1}.$$

Wallis went on to assert that this claim remained true even when the index is irrational. He looked at one such example, that of an index equal to $\sqrt{3}$.

In many cases, Wallis had no way to directly verify the characteristic ratio of an index, for example: $y = \sqrt[3]{x^2}$. It is here that he invokes his principle of "interpolation." He coined this term from the Latin root "to polish." He claimed that whenever one can discern a pattern of any kind in a sequence of examples one has the right to apply that pattern to any intermediate values if possible. That is to say that one should always attempt to polish in between. Nunn [23] calls this his principle of continuity, and claims that this is a major step towards the development of a theory of continuous functions. Its influence on Newton was profound.

Wallis did proceed boldly with his principle of interpolation, but he always sought some way to double check his patterned conjectures through an interpretation outside of his original representation. It is this confirmation through and across representations that made his interpolations so compelling. He tried to construct a continuous theory which would connect all the little islands of accepted truth. With this in mind let us look at a subsequent section of the *Arithmetica Infinitorum*, which contains his most famous interpolation.

How can we determine the characteristic ratio of the circle? This is the question that motivated Wallis to study a particular family of curves from which he could interpolate the value for the circle. He wrote the equation of the circle of radius r, as $y = \sqrt{r^2 - x^2}$, and considered it in the first quadrant. He wanted to determine the ratio of its area to the r by r square that contains it. Of course he knew that this ratio is $\pi/4$ from various geometric constructions going back to Archimedes, but he wanted to test his theory of index, characteristic ratio and interpolation by arriving at this result in a new way. This would provide a confirmation of the theory through coordination of representations.

As it stands the equation of the circle does not yield to the methods he has developed thus far. Every student of calculus confronts this when he/she finds that the general power rule will not integrate the circle. Wallis searched for a family of equations in which he could embed the circle equation. He considered the family of curves defined by the equations $y = (\sqrt[q]{r} - \sqrt[q]{x})^p$. This family is binomial, symmetric, and includes the equation of a circle as a special case $(p = q = 1/2)$.[6]

Figure 7 shows the graphs of Wallis's equations in the unit square $(r = 1)$ for $p = 1/2, q = 0, 1/2, 1, 3/2, \ldots, 5$ and for $q = 1/2, p = 0, 1/2, 1, 3/2, \ldots, 5$. The line $y = x$ has been added to display the symmetry.

If p and q are both integers, he knew that by expanding $(\sqrt[q]{r} - \sqrt[q]{x})^p$ and using his rule for characteristic ratios he could determine the ratio for these curves. Figure 8 shows this as the ratio of the shaded area to the rectangle which encloses

---

[6]The symmetry of this family of curves can be seen by rewriting the equations in the form $y^{1/p} + x^{1/q} = r^{1/q}$. Reversing x and y has a similar effect to reversing p and q. This form also displays their relation to the equation of the circle in the form $y^2 + x^2 = r^2$.

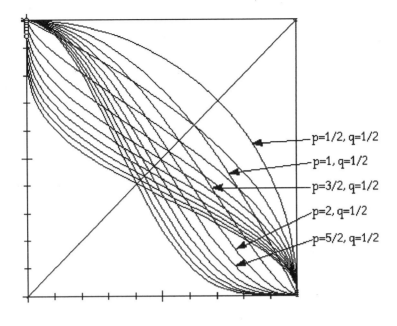

FIGURE 7

it. Note that all of these curves pass through $(r, 0)$, and that the height of the rectangle is $\sqrt[q]{r^p}$.

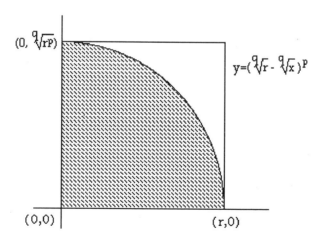

FIGURE 8

The circle can be seen as the case when $p = 1/2$, $q = 1/2$. Note here that Wallis is interpreting $\sqrt[1/2]{x}$ as x^2. He also allowed p or q to be zero. He interpreted x^0 as 1 based on his characteristic ratio rule. Since $y = x^0$ must have a characteristic ratio of 1, it must be a horizontal line. Since 1 to any power is 1 this horizontal line must be at height 1.

Wallis then calculated the characteristic ratio when p and q are both integers. For example, if $p = q = 2$, then:

$$y = \left(\sqrt{r} - \sqrt{x}\right)^2 = r - 2\sqrt{r}\sqrt{x} + x,$$

and so must have a characteristic ratio of

$$1 - 2 \cdot \frac{2}{3} + \frac{1}{2} = \frac{1}{6}.$$

Several things must be noted about this calculation. First, Wallis used the sequence of x values: $\{0, 1, 2, 3, \ldots, r\}$, and then let r increase. He assumed that the calculation was valid for any arithmetic sequence. Hence his characteristic ratio is independent of the value of r.[7] This he confirmed empirically by computing a vast array of examples. Second, when looking at the middle term, it may appear at first glance that since the characteristic ratio of $\sqrt{x}$ is $2/3$, that the -2 comes along for the ride in a linear fashion. But as we said before, characteristic ratios are not linear. The characteristic ratio of $2\sqrt{x}$ is still $2/3$. However, in this case the calculation is valid since the maximum value of the entire curve is determined by the constant term r. Since all of the values in the denominator are r, the coefficient of the middle term does not factor out of the denominator, and hence does not cancel out.

After computing the characteristic ratios when p and q are integers, Wallis noted that they were all unit fractions and that the denominators were the figurate or binomial numbers. These numbers had been known since ancient times and had recently been discussed by Pascal and others in the seventeenth century. Wallis then inverted these ratios so that they became integers and made a table of them. Table 1 records the ratio of the rectangle to the shaded area (see Figure 8) for each of the curves $y = \left(\sqrt[q]{r} - \sqrt[q]{x}\right)^p$.

At this point Wallis temporarily abandoned both the geometric and algebraic representations and began to work solely in the table representation. The question then became, how does one interpolate the missing values in this table?[8]

Wallis proceeded as follows. He first worked on the rows with integer values of q. When $q = 0$, we see the constant value of one, so fill it in with ones. When $q = 1$, we see an arithmetic progression, so fill it in with arithmetic means. In the row $q = 2$ we have the triangular numbers which are the sums of the integers in the row $q = 1$. Hence he could use the formula for the sum of consecutive integers,

$$\frac{s^2 + s}{2},$$

---

[7]One could regard "r" as the scaling factor for the sytem of coordinates which are being imposed upon a preexisting curve.

[8]At this point the reader may wish to try his/her own hand at completing this table. If these ideas were to be used with students this could generate a fascinating discussion.

$p =$ 0	$\frac{1}{2}$	1	$\frac{3}{2}$	2	$\frac{5}{2}$	3	$\frac{7}{2}$	4
$q =$								
0 1		1		1		1		1
$\frac{1}{2}$								
1 1		2		3		4		5
$\frac{3}{2}$								
2 1		3		6		10		15
$\frac{5}{2}$								
3 1		4		10		20		35
$\frac{7}{2}$								
4 1		5		15		35		70

TABLE 1

where $s = p + 1$. This formula appeared in the margin of his table. Putting the intermediate values into this formula allows us to complete the row $q = 2$. For example letting $s = 3/2$ in this formula yields 15/8, which becomes the entry where $p = 1/2$ and $q = 2$.

The numbers in the row $q = 3$ are the pyramidal numbers each of which is the sum of the integers in the row $q = 2$. Hence the appropriate formula is found by summing the formula from the row $q = 2$. Applying Alhazen's first two formulas and then collecting terms Wallis obtained

$$\frac{1}{2}\sum_{1}^{s}(i^2 + i) = \frac{s^3 + 3s^2 + 2s}{6},$$

which appears in the margin of his row $q = 3$. Letting $s = 3/2$, for example, the formula yields 105/48, which becomes the table entry where $p = 1/2$ and $q = 3$ (again $s = p + 1$).

In a similar fashion Wallis summed the previous cubic formula to obtain a formula for the row $q = 4$. Using the first three of Alhazen's formulas and then collecting terms we obtain:

$$\frac{s^4 + 6s^3 + 11s^2 + 6s}{24}.$$

This procedure completes rows where q is an integer by applying the formulas to the intermediate values. Since the table is symmetrical this also allows us to fill in the corresponding columns when p is an integer. Wallis then created a table with his formulas for interpolation written in the margin (see Table 2).[9]

$p =$ 0	$\frac{1}{2}$	1	$\frac{3}{2}$	2	$\frac{5}{2}$	3	$\frac{7}{2}$	4	
$q =$									
0 **1**	1	**1**	1	**1**	1	**1**	1	**1**	1
$\frac{1}{2}$ 1		$\frac{3}{2}$		$\frac{15}{8}$		$\frac{105}{48}$		$\frac{945}{384}$	
1 **1**	$\frac{3}{2}$	**2**	$\frac{5}{2}$	**3**	$\frac{7}{2}$	**4**	$\frac{9}{2}$	**5**	s
$\frac{3}{2}$ 1		$\frac{5}{2}$		$\frac{35}{8}$		$\frac{315}{48}$		$\frac{3465}{384}$	
2 **1**	$\frac{15}{8}$	**3**	$\frac{35}{8}$	**6**	$\frac{63}{8}$	**10**	$\frac{99}{8}$	**15**	$\frac{s^2+s}{2}$
$\frac{5}{2}$ 1		$\frac{7}{2}$		$\frac{63}{8}$		$\frac{693}{48}$		$\frac{9009}{384}$	
3 **1**	$\frac{105}{48}$	**4**	$\frac{315}{48}$	**10**	$\frac{693}{48}$	**20**	$\frac{1287}{48}$	**35**	$\frac{s^3+3s^2+2s}{6}$
$\frac{7}{2}$ 1		$\frac{9}{2}$		$\frac{99}{8}$		$\frac{1287}{48}$		$\frac{19305}{384}$	
4 **1**	$\frac{945}{384}$	**5**	$\frac{3465}{384}$	**15**	$\frac{9009}{384}$	**35**	$\frac{19305}{384}$	**70**	$\frac{s^4+6s^3+11s^2+6s}{24}$

TABLE 2

The more familiar rule for the formation of a binomial table tells us that each entry should be the sum of two others, one of which appears here two spaces up, and the other two spaces to the left. One can now check that the new interpolated values also conform to this rule of formation. Note here that the 15/8 occurs in the place $p = 2$, $q = 1/2$. The area under this curve involves exactly the same summation that occurs in the calculation of the volume of Alhazen's paraboloid, and the interpolation is consistent with Alhazen's result.

With these entries now in place, Wallis turned his attention to the row $q = 1/2$. Each of the entries that now appear there are calculated by using each of the successive interpolation formulas that appear in the margins. Each of these

---

[9]To view the original table of Wallis, see Wallis [28, p. 458] or Struik [26, p. 252].

formulas has a higher algebraic degree. What pattern exists in the formation of these numbers which will allow us to interpolate between them to find the missing entries?[10] Remember that the first missing entry is the $q = p = 1/2$ (i.e. the ratio of the square to the area of the quarter circle), and so if our table manipulations are to be validated back in the geometric representation this value should come out to be $4/\pi$.

$\frac{1}{2}$	1		$\frac{3}{2}$		$\frac{15}{8}$		$\frac{105}{48}$		$\frac{945}{384}$

At first Wallis tried to fill in this row with arithmetic averages. The average of 1 and $3/2$ is $5/4$. But if $4/\pi$ is equal to $5/4$, then $\pi = 3.2$, which is not quite correct. If geometric averaging is attempted, the value for π comes out even larger (approximately 3.266).

Wallis now observed that each of the numerators in these fractions is the product of consecutive odd integers, while each of the denominators is the product of consecutive even integers. That is to say,

$$\frac{15}{8} = \frac{3 \cdot 5}{2 \cdot 4}, \quad \frac{105}{48} = \frac{3 \cdot 5 \cdot 7}{2 \cdot 4 \cdot 6}, \quad \text{and} \quad \frac{945}{384} = \frac{3 \cdot 5 \cdot 7 \cdot 9}{2 \cdot 4 \cdot 6 \cdot 8}.$$

Hence to move two entries to the right in this row one multiplies by $\frac{n}{n-1}$. For the entries that appear so far, n is always odd. So Wallis assumed that to get from one missing entry to the next one he should still multiply by $\frac{n}{n-1}$, but this time n would have to be the intermediate even number. Denoting the first missing entry by Ω,[11] then the next missing entry should be $\frac{4}{3}\Omega$, and the one after that should be $\frac{4 \cdot 6}{3 \cdot 5}\Omega = \frac{8}{5}\Omega$, and the one after that should be $\frac{4 \cdot 6 \cdot 8}{3 \cdot 5 \cdot 7}\Omega = \frac{64}{35}\Omega$, and so forth. The $q = 1/2$ row now becomes:

$\frac{1}{2}$	1	Ω	$\frac{3}{2}$	$\frac{4}{3}\Omega$	$\frac{15}{8}$	$\frac{8}{5}\Omega$	$\frac{105}{48}$	$\frac{64}{35}\Omega$	$\frac{945}{384}$

The column $p = 1/2$ can now be filled in by symmetry. The row $q = 3/2$ has a similar pattern (i.e. products of consecutive odd over consecutive even numbers) but there the entries two spaces to the right are always multiplied by $\frac{n}{n-3}$. One can also double check, as Wallis did, that this law of formation agrees with the usual law for the formation of binomials, i.e. each entry is the sum of the entries two up and two to the left. Filling in the rest of the entries in this manner results in Table 3.

---

[10]Once again, we would encourage the reader to stop and try his/her own hand at filling in the missing entries.

[11]Wallis used a small square to denote this missing entry.

$p =$	0	$\frac{1}{2}$	1	$\frac{3}{2}$	2	$\frac{5}{2}$	3	$\frac{7}{2}$	4	
$q =$										
0	**1**	1	**1**	1	**1**	1	**1**	1	**1**	1
$\frac{1}{2}$	1	Ω	$\frac{3}{2}$	$\frac{4}{3}\Omega$	$\frac{15}{8}$	$\frac{8}{5}\Omega$	$\frac{105}{48}$	$\frac{64}{35}\Omega$	$\frac{945}{384}$	
1	**1**	$\frac{3}{2}$	**2**	$\frac{5}{2}$	**3**	$\frac{7}{2}$	**4**	$\frac{9}{2}$	**5**	s
$\frac{3}{2}$	1	$\frac{4}{3}\Omega$	$\frac{5}{2}$	$\frac{8}{3}\Omega$	$\frac{35}{8}$	$\frac{64}{15}\Omega$	$\frac{315}{48}$	$\frac{128}{21}\Omega$	$\frac{3465}{384}$	
2	**1**	$\frac{15}{8}$	**3**	$\frac{35}{8}$	**6**	$\frac{63}{8}$	**10**	$\frac{99}{8}$	**15**	$\frac{s^2+s}{2}$
$\frac{5}{2}$	1	$\frac{8}{5}\Omega$	$\frac{7}{2}$	$\frac{64}{15}\Omega$	$\frac{63}{8}$	$\frac{128}{15}\Omega$	$\frac{693}{48}$	$\frac{512}{35}\Omega$	$\frac{9009}{384}$	
3	**1**	$\frac{105}{48}$	**4**	$\frac{315}{48}$	**10**	$\frac{693}{48}$	**20**	$\frac{1287}{48}$	**35**	$\frac{s^3+3s^2+2s}{6}$
$\frac{7}{2}$	1	$\frac{64}{35}\Omega$	$\frac{9}{2}$	$\frac{128}{21}\Omega$	$\frac{99}{8}$	$\frac{512}{35}\Omega$	$\frac{1287}{48}$	$\frac{1024}{35}\Omega$	$\frac{19305}{384}$	
4	**1**	$\frac{945}{384}$	**5**	$\frac{3465}{384}$	**15**	$\frac{9009}{384}$	**35**	$\frac{19305}{384}$	**70**	$\frac{s^4+6s^3+11s^2+6s}{24}$

TABLE 3

The table is now complete except for the determination of the value of Ω. One might be tempted to think that we are done, since we know from geometric considerations that Ω must equal $4/\pi$. Wallis already knew from Archimedes one way to evaluate π by inscribing polygons in the circle. This ancient method of calculating π leads, algebraically, to a series of nested square roots, one for each doubling of the number of sides in the polygon. Using this geometric result at this point would, however, violate the whole program of Wallis. He had to find a way to calculate Ω using his principle of interpolation so that he could check his value against the one known from geometry. It is only in this way that he created a critical experiment capable of testing and validating his method of interpolation.[12]

So returning once again to the row $q = 1/2$ where moving two spaces to the right from the nth entry multiplies that entry by $\frac{n}{n-1}$ Wallis noted that as n increases the fraction $\frac{n}{n-1}$ gets closer and closer to 1. Hence the number two spaces to the right must change very little as we go further out the sequence.

---

[12]The philosopher Thomas Hobbes criticized the work of Wallis as a "scab of symbols." Hobbes saw no reason why the results of algebra should be consistent with those of geometry. The response of Wallis to these comments was a great embarrassment to the aging philosopher (Hobbes). See Cajori [5].

This is true of the calculated fractions as well as the multiples of Ω Wallis argued that since the whole sequence is monotonically increasing, that consecutive terms must also be getting close to one another as we proceed. Hence as we build these terms we should have that:

$$\Omega \frac{4 \cdot 6 \cdot 8 \cdot 10 \ldots}{3 \cdot 5 \cdot 7 \cdot 9 \ldots} \approx \frac{3 \cdot 5 \cdot 7 \cdot 9 \ldots}{2 \cdot 4 \cdot 6 \cdot 8 \ldots}$$

$$\Omega \approx \frac{3 \cdot 3 \cdot 5 \cdot 5 \cdot 7 \cdot 7 \cdot 9 \cdot 9 \ldots}{2 \cdot 4 \cdot 4 \cdot 6 \cdot 6 \cdot 8 \cdot 8 \cdot 10 \ldots}.$$

Since Ω should be equal to $4/\pi$ Wallis's interpolations are justified provided that:

$$\pi \approx 2 \cdot \frac{2 \cdot 2 \cdot 4 \cdot 4 \cdot 6 \cdot 6 \cdot 8 \cdot 8 \ldots}{1 \cdot 3 \cdot 3 \cdot 5 \cdot 5 \cdot 7 \cdot 7 \cdot 9 \ldots}.$$

If one calculates this infinite product it does indeed converge to π. That is it agrees with the results of Archimedes and others who had previously calculated π within a geometric representation. It is this final coordination of multiple representations that confirmed for Wallis the viability of his principle of interpolation. This was an entirely new way to calculate π, and it became famous. It is often mentioned in modern analysis books and its connections to other topics are discussed, but rarely does anyone mention its epistemological significance. This calculation is a critical empirical experiment which confirms the consistency of geometry and arithmetic. This, as Wallis would say, has been "studiously concealed" by logical analysis. Wallis wanted his work to convey the usefulness of empirical methods of investigation within mathematics. To just see this result as a new way to calculate π is to "wholly mistake the design of the treatise" (Nunn [23, p. 385]).

The empirical methods of Wallis led the young Isaac Newton to his first profound mathematical creation: the expansion of functions in binomial series (Whiteside [30]). The interested reader should look at Newton's annotations to Wallis contained in his early notebooks to see how far this method of interpolation can lead [22]. Wallis's method of interpolation became for Newton the basis of his notion of continuity. Newton generalized the methods of Wallis, but for several generations, the epistemology used by Wallis remained the dominant force in mathematics. The elaborate investigations of Euler in the eighteenth century, which included series expansions over complex numbers and the solution of many differential equations, nonetheless remained rooted in an empirical epistemology of multiple representations (Euler [17]).

Conclusions

The methods of investigation outlined in this paper have been largely purged from our mathematics curriculum. These mathematical results are now presented to students in a formal logical setting that came about in the nineteenth and twentieth centuries. They are presented as a done deed, a finished object.

Almost no sense of the activity of investigation remains. Both for the mathematical specialist and the non-specialist this type of presentation is dominant in our mathematics classrooms.

The practice of conjecture by analogy and the use of informal induction combined with coordination of multiple representations would greatly invigorate our teaching practices. For the non-specialist many practical methods of mathematical reasoning are entirely concealed. So many results of mathematics could be presented much earlier and in a simpler setting if empirical methods were encouraged. Even for the student who chooses to specialize in mathematics most of the process of invention is never highlighted, though much of the work of a professional mathematician is carried out in this way.

The availability of calculators, and computers make it possible for many people to engage in empirical speculations. Indeed many of the most interesting area of modern research in mathematics are being carried out as computer experiments. The fractal geometry developed by Mandelbrot and others is a good example. Astounding new pictures have been constructed that involve little formal analysis. These empirical methods are now being applied profoundly in biology. Recently some of the most original mathematical research has come from outside of math departments.

In summary we would say that it is the mathematical action rather than the result which most deeply conveys meaning. It is the construction of a mathematical setting where these actions can be checked across different representations that produces confidence in the viability of a method. It is the understanding of method that empowers students. Logical analysis can be very beautiful and satisfying in its own way, but it is like a spider spinning its webs in the castle of mathematics. The danger is that after a while one begins to believe that the webs hold up the castle.[13] Students exposed only to traditional, overly formal curricula are especially prone to this danger.

REFERENCES

1. Baron, M., *The Origins of Infinitesimal Calculus*, Pergamon, London, 1969.
2. Boyer, C.B., *A History of Mathematics*, Wiley, New York, 1968.
3. Boyer, C.B., *History of Analytic Geometry*, Chapters III–V, Scripta Mathematica, New York, 1956.
4. Boyer, C.B., *Pascal's formula for the sums of powers of integers*, Scripta Mathematica **9** (1943), 237–244.
5. Cajori, F., *History of exponential and logarithmic concepts*, American Mathematical Monthly **20** (1913).
6. Cajori, F., *Controversies on mathematics between Wallis, Hobbes, and Barrow*, The Mathematics Teacher **XXII(3)** (1929), 146–151.

---

[13] "The developments in this century bearing on the foundations of mathematics are best summarized in a story. On the banks of the Rhine, a beautiful castle had been standing for centries. In the cellar of the castle, an intricate network of webbing had been constructed by industrious spiders who lived there. One day a strong wind sprang up and destroyed the web. Frantically the spiders worked to repair the damage. They thought it was their webbing that was holding up the castle." (from Morris Kline's *Mathematics, the Loss of Certainty*, p. 277).

7. Calinger, R., *Classics of Mathematics*, Moore, Oak Park, IL, 1982.

8. Confrey, J., *Multiplication and splitting: their role in understanding exponential functions*, Proceedings of the Tenth Annual Meeting of the North American Chapter of the International Group for the Psychology of Mathematics Education (PME-NA), DeKalb, IL, 1988.

9. Confrey, J., *Using computers to promote students' inventions on the function concept*, This Year in School Science 1991 (Malcom, Roberts, and Sheingold, eds.), American Association for the Advancement of Science, Washington DC, 1992, pp. 131–161.

10. Confrey, J., *Learning to see children's mathematics: crucial challenges in constructivist reform*, The Practice of Constructivism in Science Education (Tobin, K., ed.), American Association for the Advancement of Science, Washington DC, 1993, pp. 299–321.

11. Confrey, J., *Splitting , similarity, and rate of change: New approaches to multiplication and exponential functions*, The Development of Multiplicative Reasoning in the Learning of Mathematics (Harel, G. and Confrey, J., eds.), State University of New York Press, Albany NY, 1994, pp. 293–332.

12. Confrey, J. and Smith, E., *Applying an epistemology of multiple representations to historical analysis: A review of "democratizing the access to calculus: New routes to old roots" by James Kaput*, Mathematical Thinking and Problem Solving (Schoenfeld, A., ed.), Lawrence Erlbaum Assoc. Inc., Hillsdale NJ (to appear).

13. Courant, R., *Gauss and the present situation of the exact sciences*, Mathematics: People, Problems, Results (Campbell, D. and Higgins, J., ed.), vol. I, Wadsworth International, Belmont CA, 1984, pp. 125-133.

14. Dennis, D., *Historical perspectives for the reform of mathematics curriculum: geometric curve drawing devices and their role in the transition to an algebraic description of functions*, (unpublished doctoral dissertation), Cornell University, Ithaca, NY, 1995.

15. Descartes, R., *The Geometry*, (Translated by D.E. Smith and M.L. Latham), Open Court, LaSalle, IL, 1952.

16. Edwards, C.H., *The Historical Development of Calculus*, Springer-Verlag, New York, 1979.

17. Euler, L., *Introduction to Analysis of the Infinite*, (Translated by J. D. Blanton), Springer-Verlag, New York, 1988.

18. Fowler, D. H., *The Mathematics of Plato's Academy: A New Reconstruction*, Clarendon Press, Oxford, 1987.

19. Katz, V., *A History of Mathematics: An Introduction*, Harper Collins, New York, 1993.

20. Lakatos, I., *Proofs and Refutations, The Logic of Mathematical Discovery*, Cambridge University Press, New York, 1976.

21. Mahoney, M. S., *The Mathematical Career of Pierre de Fermat*, Princeton University Press, Princeton NJ, 1973.

22. Newton, I., *The Mathematical Papers of Isaac Newton, Volume I: 1664-1666*, Cambridge University Press, Cambridge, 1967.

23. Nunn, T. P., *The arithmetic of infinities*, Mathematics Gazette **5** (1909-1911), 345–356 and 377–386.

24. Scott, J. F., *The Mathematical Work of John Wallis*, Chelsea Publishing Co., New York, 1981.

25. Smith, E., Dennis, D. and Confrey, J., *Rethinking functions, cartesian constructions*, The History and Philosophy of Science in Science Education, Proceedings of the Second International Conference on the History and Philosophy of Science and Science Education (S. Hills, ed.), The Mathematics, Science, Technology and Teacher Education Group; Queens University, Kingston, Ontario, 1992.

26. Struik, D.J., *A Source Book in Mathematics:1200–1800*, Harvard University Press, Cambridge MA, 1969.

27. Unguru, S., *On the need to rewrite the history of Greek mathematics*, Archive for History of Exact Sciences **15(2)** (1976), 67–114.

28. Wallis, J., *Arithmetica Infinitorum*, Opera Mathematica, vol. 1, Georg Olms Verlag, New York, 1972, pp. 355–478.

29. Wittgenstein, L., *Remarks on the Foundations of Mathematics*, (translated by G.E.M. Anscombe), MIT Press, Cambridge MA, 1967.

30. Whiteside, D. T., *Newton's discovery of the general binomial theorem,* Mathematics Gazette
    45 (1961), 175–180.
31. Whiteside, D. T., *Patterns of mathematical thought in the later 17th century,* Archive for
    History of Exact Sciences, vol. 1, 1960–1962, pp. 179–388.

Appendix—Negative Exponents and Ratios

We have often found it interesting to examine some of the ideas in mathematics that did not gain general acceptance. The serious consideration of these alternative conceptions can enlighten our thinking and our teaching practice as we try to understand student conceptions. The following examination of Wallis' use of negative values within his theory of index and ratio is a good example.

Wallis interprets negative numbers as exponents in the same way that we do. That is, he defines the index of $1/x$ as -1, the index of $1/x^2$ as -2, and so on. He also extends this definition to fractions: for example, $1/\sqrt{x}$ has an index of $-1/2$. He then claims that the relationship between the index and the characteristic ratio is still valid for these negative indices. That is, if k is the index then $1/(k+1)$ is the ratio of the shaded area under the curve to the rectangle (see Figure 9). In the case of a negative index this shaded area is unbounded. This does not deter Wallis from generalizing his claim.

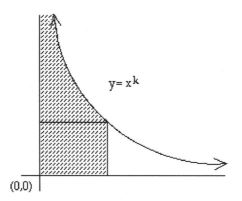

FIGURE 9

When $k = -1/2$, the characteristic ratio should be

$$\frac{1}{\left(-\frac{1}{2}+1\right)} = 2.$$

This value is indeed correct, for the unbounded area under the curve $y = 1/\sqrt{x}$ does converge to twice the area of the rectangle. This is true no matter what right hand endpoint is chosen.

When $k = -1$, the characteristic ratio should be

$$\frac{1}{-1+1} = \frac{1}{0} = \infty$$

(Wallis introduced this symbol for infinity into mathematics). Wallis accepted this ratio as reasonable since the area under the curve $y = 1/x$ diverges. This can be seen from the divergence of the harmonic series

$$1 + \frac{1}{2} + \frac{1}{3} + \frac{1}{4} + \cdots = \infty,$$

which had been known since at least the fourteenth century (Boyer [2, Chapter XIV]).

When $k = -2$, the characteristic ratio should be

$$\frac{1}{-2+1} = \frac{1}{-1}.$$

Here Wallis's conception of ratio differs from our modern arithmetic of negative numbers. He does not believe that $1/(-1) = -1$. Instead he stays with his epistemology of multiple representations. Since the shaded area under the curve $y = 1/x^2$, is greater than the area under the curve $y = 1/x$, he concludes that the ratio $1/(-1)$ is greater than infinity ("ratio plusquam infinita") (Nunn [23, p. 355]). He goes on to conclude that $1/(-2)$ is even greater. This explains the plural in the title of his treatise *Arithmetica Infinitorum*. The appropriate translation would be *The Arithmetic of Infinities*.

Most historians of mathematics quickly brush over this concept if they mention it at all. Those who mention it quickly cite the comments of the French mathematician Varignon (1654–1722), who pointed out that if the minus sign is dropped in the ratio then we arrive at the correct ratio of the unshaded area under the curve to the area of the rectangle. This was an instance of the beginning of the idea that negative numbers could be viewed as complements or reversals of direction.

We, however, find it well worth pondering Wallis' original conception. In what ways does it make sense to consider the ratio of a positive to a negative number as greater than infinity? In the area interpretation from Figure 9, we could view these different infinities as greater and greater rates of divergence. Such views are often taken in mathematics. The area under $y = 1/x^3$ does diverge faster than the area under $y = 1/x^2$.

Let's consider an even simpler situation. If I have 1 dollar, and you have 50 cents, then we say that I have twice as much money as you. If I have 1 dollar, and you have 10 cents then we say that I have ten times as much money as you. If I have 1 dollar, and you have nothing, then we could say that I have infinitely more money than you. Many mathematicians would accept this statement. Now if I have 1 dollar, and you are in debt, shouldn't we say that the ratio of my money to yours is even greater than infinity? This seems to us to be a question that is worth pondering.

CORNELL UNIVERSITY, ITHACA, NEW YORK

CBMS Issues in Mathematics Education
Volume **6**, 1996

Dihedral Groups: A Tale of Two Interpretations

RINA ZAZKIS AND ED DUBINSKY

This note is concerned with mathematical objects and their naming schemes, that is, means of assigning to an object some algebraic quantity. We start with an empirically-based examination of a phenomenon observed in undergraduate Abstract Algebra courses and then we continue with the analysis of related phenomena that are found in the literature.

Our specific interest is with mathematical and psychological aspects of constructing a group on the set of symmetries of a regular polygon of n sides and on the set of permutations of n objects. The points we wish to make are amply illustrated in a specific example and so we will mainly consider the case $n = 4$ with the corresponding set D_4 of 8 symmetries, and the symmetric group S_4 of 24 permutations. As is well known, the elements of D_4 correspond to elements of S_4 and so it would seem that there are two ways of constructing (representing, if you like) a group structure on D_4: one as motions of the square with composition (of transformations) as the operation and the other as a set of permutations with multiplication (of permutations) as the operation. They should be isomorphic.

One can see the first construction as a geometric or visual way of thinking about a certain group and the second as a symbolic or analytic way of thinking of the same group. We were (and are) interested in how students used these two modes of thinking when they were trying to understand the construction of these groups. We found that students tended to make a certain error in moving back and forth between the two representations and, in analyzing this error we came to the view that the situation was a little more complicated (both mathematically and psychologically) than one might think. Indeed, we even came to question our belief that the group D_4 can be constructed in a completely visual manner.

We think that these questions are important because they are related to the pedagogical issue of students' visual and analytic approaches to making sense out of mathematical situations. The observations we are making in this note form a small piece of a larger study of general visual/analytic issues [11].

In the following pages, we explain how we were motivated to think through the issues discussed in this paper, state the main problem, propose an explanation of the students' difficulty, suggest an analysis that could form the basis for eliminating the difficulty, and describe how we think students may be thinking about the relationship between dihedral groups and groups of symmetries.

Having done that, we attempt to clarify certain complex relationships between mathematical objects and their names. We analyze several related problems (of double interpretation) described in literature, such as Birkhoff and Mac Lane's alibi/alias dichotomy [3, 9]. Our main purpose in such an analysis is to reveal complexities that are often not acknowledged by mathematics experts. We explain in what way our problem is related to other problems from the literature and in what way it differs. We also suggest possible avenues for future research that will further explore the mathematical and psychological phenomena we have encountered.

Our Motivation

Our considerations were motivated by ten individual interviews with undergraduate mathematics majors in the middle of a first course in Abstract Algebra. In one of the tasks presented to the students in the interview, they were asked to list the elements of D_4, and then to calculate a product of two specific elements of D_4. The interviewer made no attempt to suggest a particular representation of D_4 and each student made one of two choices of how to do it: geometrically, using a physical model of a square, or analytically, multiplying permutations. The interviewer then asked the student for another way to do it and to see that both methods gave the same answer. As it turned out, eight of the ten students did not get the same answer and felt they had made an error (they each did the same thing) and this note resulted from our attempts to understand what they did and why it seems wrong.

Let us introduce the notation used in the Abstract Algebra course which these students were taking. The elements of D_4 were familiar to them as four clockwise rotations, denoted R_0, R_{90}, R_{180}, R_{270}, and four reflections across horizontal, vertical and diagonal axes, denoted H, V, D_L (left diagonal) and D_R (right diagonal). (See Figure 1.) The permutations of S_4 were denoted as sequences of four digits, that represent the "first floor" of a standard "double decker" notation. For example, [4123] represented the permutation "$1 \to 4$, $2 \to 1$, $3 \to 2$, $4 \to 3$," usually listed as $\binom{1234}{4123}$.

The following excerpt from an interview with Peter[1] was typical of the student performance of the task of computing $R_{90} * V$ and their reactions upon its completion. Note that the student applies the convention of starting with the element on the right in computing a product, that is, $R_{90} * V$ is interpreted as "V followed by R_{90}." (See Figure 2.)

---

[1]The names of the interviewees have been changed to protect their identities.

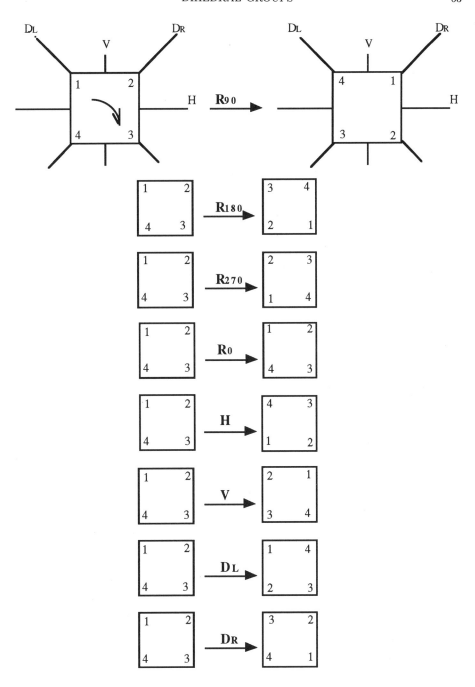

FIGURE 1. Elements of D_4—Global interpretation.

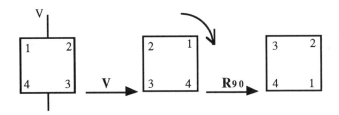

FIGURE 2.   Carrying out V followed by R_{90}.

Peter: $R_{90} * V$? I would take this, do the vertical flip, and then do R_{90}, which is a 90 degree rotation to the East, and that would give me [3214].

Int: And according to your element table there, which element would that be?

Peter: [3214] is D_R.

Int: Now let's do it again and this time with permutations.

Peter: OK, $R_{90} * V$. R_{90} from our permutations is [4123] and V is [2143]. OK, and the way I would do it was start and feed 1 into 2 and then . . . , 2 in R_{90} feeds into 1, so 1 goes to 1. Uh, 2 goes to 1, 1 goes to 4; 3 goes to 4, 4 goes to 3; and 4 goes to 3, 3 goes to 2. [1432] which is D_L.

Int: Ummm, that's interesting.

Peter: Yes, it is. (chuckles softly).

Int: Why do you think we got different answers? (pause) More than likely, they're not both correct.

Peter: I'd say they're not. D_L is certainly not D_R. So what did we do wrong? We did a V and then R_{90}, [3214]. Did I put them down wrong? R_{90} is [4123] and V is [2143], so I didn't do that wrong. (pause) I'm stumped (chuckle).

Peter appears to perform the manipulations and calculations correctly, but using two different methods, he gets two different answers. He sees this and, in listening to the tape we took his reaction to be a kind of nervous laughter indicating some discomfort with the situation or what is called in learning theory, disequilibration [4]. Since the calculations are performed correctly, the difficulty appears to lie in the two representations and their connections.

The Problem

The conventional view is that the dihedral groups D_n are established from geometrical considerations of rigid motions of a regular polygon of n sides and the symmetric groups S_n are established from considerations of permutations of n objects. Then, there is an isomorphism of D_n into S_n. Textbooks on group theory, however, usually don't spell out these isomorphisms to the learner explicitly. The books prefer to leave this as an exercise to the reader (e.g., Armstrong [2, p. 36] or to list the elements of D_3 or D_4 as permutations, without making explicit correspondence to the transformations they represent (e.g., Fraleigh [6, p. 70]. As demonstrated in the previous section, learners who try to use an explicit correspondence, can be surprised by unexpected results. This raises pedagogical and attitudinal issues. Will students try to resolve these contradictions or will they give up and eliminate their disequilibration by making unnecessary conclusions regarding the nature of mathematical anomalies? What pedagogical strategies are promising for getting more of the former and less of the latter?

In trying to analyze the mathematical situation from a psychological point of view, we tried to understand how one might think about the connection between D_4 and S_4. Following the traditional point of view we began with introspection into our own way of constructing the group D_4. The idea was to do this first and then think about the connection with S_4.

The eight symmetries of a square are clear. They are dynamic processes which with a little mental activity that is non-trivial, but reasonable for the students we have in mind, can be thought of as eight individual objects and it is reasonable to name them as in Figure 1. Do these eight objects form a group? Previous research has reported [5] that beginners in Group Theory have a tendency to ignore the operation and relate only to the set of elements of the group. However, even when the need for operation is recognized, applying the binary operation of composition to the symmetries of a square is not as obvious as it may seem. To avoid the undue influence of permutations we tried to do this without labeling the vertices. It is not so easy, for example, for the composition $R_{270} * H$. That got us to wonder about the "easy" cases, say $R_{270} * R_{180}$. Obviously the answer is R_{90}. But is it?

Take a square which is completely blank and looks exactly the same on both sides. Rotate it by 180 degrees and then by 270 degrees. In what sense can one say that you have rotated the square 90 degrees? You can't tell by the result, because the square looks exactly like it would look if you had not rotated it at all. You can't tell by watching the action because, in reality the composition of the two acts of rotation is indeed an act of rotation—of 450 degrees, not of 90 degrees! A combination of more thought and choice of convention has to go into a decision to call it rotation by 90 degrees.

In this sense, one can suggest that specifying the binary operation on a set of symmetries is not sufficient to define a group, since from a purely visual point of view, unless some analytic structure and representational conventions are introduced (labeling of vertices, addition mod 360, etc.) the eight rigid motions of a square are not closed under composition. There is a real difficulty in specifying the net result of a sequence of transformations without some form of labeling. It seems that labeling vertices implies an embedding of the set D_4 in the set S_4 and carrying the group structure to D_4 from (a subset of) S_4. Our conclusion from this is that the connection between D_n and S_n, if considered carefully, must be established on the level of sets, not as groups. One constructs a one-to-one map of D_n into S_n (using a considerable amount of visualization) and then the operation on D_n is constructed without ambiguity by pulling back the operation on S_n. It may seem that paying explicit attention to the construction of this injection could help students avoid pitfalls related to this connection.

Unfortunately, this is not quite enough. There is more than one reasonable way to embed the set D_n in the set S_n.

Two Embeddings

There are (at least) two possible ways to correspond a symmetry of a (labeled) square to a permutation. One way is to look at the vertices of the square as objects being moved. In performing R_{90}, vertex 1 is moved to vertex 2, vertex 2 is moved to vertex 3, 3 is moved to 4 and 4 is moved to 1. Therefore we may represent R_{90} as [2341]. We will refer to this correspondence as "object interpretation."

Another way of interpreting the situation is to look at the environment of the square and think not about vertices moving but about positions and which vertices they contain. In this interpretation, after R_{90}, position 1 contains vertex 4, position 2 contains vertex 1, position 3 contains 2, and 4 contains 1. This is represented by a permutation [4123].

We will refer to this correspondence as "position interpretation."

Clearly, these two correspondences are inverses in the sense that, given an element of D_4, the permutation to which it corresponds under the object interpretation is the inverse of the permutation to which it corresponds under the position interpretation. That is, there is an underlying anti-automorphism of the group S_4 and hence there are two, essentially different, ways in which groups can be constructed on the set D_4 by using permutations. The difference, of course, is small in the case of D_4 since all but two of its elements are idempotents. For example, the fact that R_{180} has order two assures that when vertex 1 moves to vertex 3, position 1 contains vertex 3; when vertex 2 moves to vertex 4, position2 contains vertex 4; and so on. The same situation exists with the reflections. We can predict from this analysis that the error which our students make will never appear unless we ask them to form a product involving R_{90} or R_{270} as either a factor or a result.

We can use our analysis to give a plausible explanation of the errors made by our students. Peter made a square, labeled the sides and performed the manipulations as indicated in Figure 2. There is no question about what to do with the first motion—the square is flipped across its vertical axis. The problem arises with the second motion because the square is no longer in its original position. In order to decide whether applying R_{90} means that the square should be rotated clockwise or counter-clockwise, it is necessary to describe the correspondence more carefully and we will do this below. We will see that a clockwise rotation is consistent with the object interpretation. The final position of the square can only be interpreted as the motion D_R and so, Peter's first response is correct.

Why did Peter get a different answer when he did it by permutation product? Notice that in choosing [4123] for R_{90} he says that "1 goes to 4," that is, position 1 contains vertex 4 which is the position interpretation. Indeed, as can be seen by comparing the excerpt with Figure 3, Peter used the position interpretation in going from symmetries to permutations, and therefore, he was in effect calculating this time with group elements that are the inverses of the V and R_{90} used in his first calculation.

Element of D4	Object interpretation	Position interpretation
R_0	[1234]	[1234]
R_{90}	**[2341]**	**[4123]**
R_{180}	[3412]	[3412]
R_{270}	**[4123]**	**[2341]**
H	[4321]	[4321]
V	[2143]	[2143]
D_L	[1432]	[1432]
D_R	[3214]	[3214]

FIGURE 3.   Corresponding transformations to permutations.

We can see explicit descriptions of how students used the position interpretation in corresponding symmetries to permutations. For example, Stacey explains how she corresponded a permutation to R_{90}.

Stacey: [1234] was my beginning position. And then I rotated it once , 90 degrees (clockwise), yes. OK, and then again by reading the top left all the way down through bottom left, my first rotation of 90 degrees was [4123].

John mentions "position" in his answer.

Int: How would you write out R_{90} as a permutation?

John: OK, um, R_{90} would map the corners [1234] to the new positions [4123].

And Jeff explicitly acknowledges the choice of "positional" interpretation in his explanation.

Int: Alright, you made a rotation 90 degrees in a clockwise fashion. Alright, now how did you get the permutation to correspond to the set?

Jeff: OK, the first element in this permutation would be where the 1 used to be. And the second one would be where the 2 used to be, the third one where the 3 used to be and the fourth one where the 4 used to be. So we got [4123] from there.

These excerpts suggest that for many students the position interpretation is natural for setting up the correspondence. However, we submit that it is not so easy to use this interpretation in actually manipulating a square. For example, consider the middle square in Figure 2. We will argue below that to be consistent with the position interpretation the correct implementation here of R_{90} is what appears to be a counterclockwise rotation. At the very least, the reader might agree at this point that for the middle square in Figure 2, the position interpretation does not tell us how to interpret R_{90}. The reason is that this interpretation involves "vertices repositioned" from an initial state. But after performing one motion, the initial positions have changed. Thus we suggest that the interpretations we have given are reasonable for deciding which permutations to assign to a single motion, but can be confusing to use in a context in which more than one motion is being performed. We need a more powerful model.

Local And Global Transformation—A Possible Solution

We would like to suggest such a model. The position interpretation corresponds to a different perspective on symmetries that we will call "local." According to this perspective, a symmetry is a "local" transformation of an object in the sense that the object is moved according to axes and directions that are transformed along with the object. The term "local," as well as the idea of local representation, is borrowed from Turtle Geometry [1, 8] where transformations are functions of the position of the object transformed.

More precisely, in the case of D_4, the idea is to construct the four axes and directions on the square and move them along with the square as shown in Figure 4.

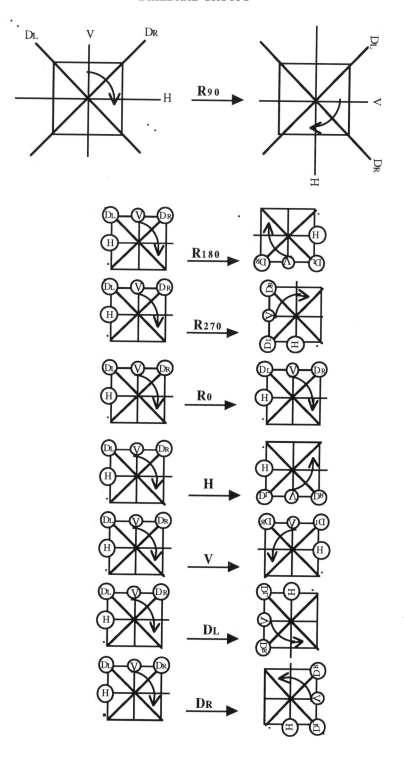

FIGURE 4. Elements of D_4—Local interpretation.

The "global" interpretation corresponds to a perspective on symmetries that is already illustrated in Figure 1. Here we consider that in each symmetry the entire plane is transformed, but the axes and directions remain fixed. The motion of the square results from this global transformation.

How can this model be used in practice? Figure 3 shows how each symmetry corresponds to a permutation in the two interpretations. The inflexible requirement is that you first choose one of the two interpretations and then you must use it both in going back and forth between symmetries and permutations as well as in deciding how to implement a composition with a square.

For example, we have seen in Figure 2 how to perform $R_{90} * V$ in the object or global interpretation. One performs the motions relative to an immutable set of axes which are fixed in the plane. As indicated in Figure 1, to perform the transformation, one does not need any information on the square, only the axes and directions on the plane that contains the square. The labels on the vertices are used only in deciding relative to these fixed axes which transformation has resulted.

In the case of the same operation performed according to the position or local interpretation, the axes and directions must be considered to be on the square as indicated in Figure 4. We choose the "canonical starting position" to be the position in which the names of the axes make sense (V is the vertical axis, D_L the left diagonal, etc.). The transformations are defined entirely in terms of these axes which are fixed on the square. Thus, the transformation V is reflection of the square about the V-axis and a rotation moves the V-axis towards the adjacent D_R-axis. In performing $V * R_{90}$ as shown in Figure 5, first R_{90} is performed on the canonical position. Then, if we want to follow up with V, we should flip the square according to the new position of the V-axes, that is, "horizontally" in the plane, since the original "vertical" axis was transformed to horizontal position by 90 degrees rotation. The resulting transformation is D_R and one does not need to keep track of the labels on the vertices to determine this. In performing $R_{90} * V$ as shown in Figure 6, we first start with V performed on the canonical position. Then, to follow up with R_{90}, we have to rotate the square *left* or counterclockwise, since the direction, which is a part of the square, has been changed by the flip. The result is D_L, which is consistent with Peter's multiplication of permutations.

Connecting The Two Models

It is perhaps reasonable to think about the object-position dichotomy as more a property of permutations and the global-local characterizations as belonging to the symmetries. In the former case, one must label the vertices and consider whether one is moving these four objects to new positions or changing what appears at a given position. For the latter case, it is the axes that are labeled and one must distinguish between moving the plane which happens to contain a square versus moving a square which happens to sit in a plane. In either case,

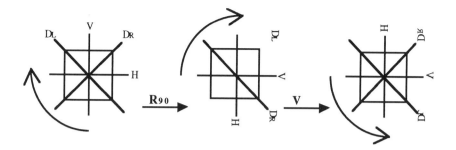

FIGURE 5. Carrying out R_{90} followed by V—Local interpretation.

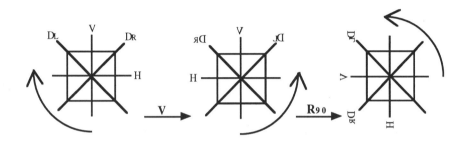

FIGURE 6. Carrying out V followed by R_{90}—Local interpretation.

some analytic structure must be added to the square. This gives further support to our earlier observation that the analytic structure of labeling vertices or axes is necessary to formally define the group of symmetries of a regular polygon with the operation of composition.

As we have indicated above, the important thing is to use a single interpretation throughout any particular discussion. This interpretation must remain invariant as one moves back and forth between permutations and symmetries or switches from listing elements to performing the group operation. The overall situation is described in Figure 7. The maps *obj* and *pos* are bijections of D_4 onto subgroups of S_4 , defined respectively by columns 1,2 of Figure 3 for *obj* and columns 1,3 of Figure 3 for *pos*. The map *inv* sends an element of S_4 into its inverse. All of the other maps in the diagram are the identity on D_4 . The entire diagram commutes. For the identity maps this is clear and for the right hand portion it follows from the relation,

$$pos(x) = inv(obj(x)) \qquad \text{for all } x \text{ in } D_4$$

The horizontal identity maps are automorphisms of D_4 with the indicated operations because the operation obtained from the object (respectively, position) interpretation is the same as the operation obtained from using global (respectively, local) axes. Another way of expressing this is that global transformations are consistent with the object interpretation and local transformations are con-sistent with the position interpretation.

The vertical identity maps are anti-isomorphisms in the sense that

$$i(xy) = i(y)i(x)$$

where i is the name of one of them.

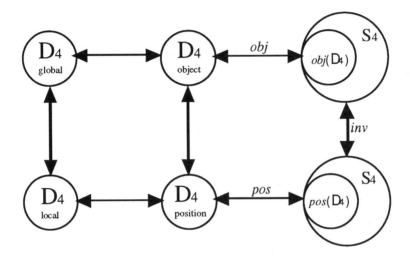

FIGURE 7.   Interpretations of binary operations on D_4.

Following through the interview segment with Peter, we suggest that he used global interpretations for performing transformations with a physical square in his first calculation, obtaining the answer D_R. In his second calculation he used the position interpretation to correspond permutations to transformations, obtaining the answer D_L. Put another way, in his first calculation he worked at the upper level of the diagram and in his second calculation he worked on the lower level. The two structures are connected by maps which do not always preserve the operations and this explains why he obtained two different answers.

Of course, in the case of commuting elements the inconsistent choice of global symmetries and position interpretation doesn't lead to an error and disequilibra-tion but leaves "an illusion of isomorphism."

Dealing With Disequilibration

As we indicated above, the error and the inconsistent interpretations which caused it were seen explicitly in eight of the ten students that were interviewed. Actually, neither of the other interviewees did the problem correctly. One stu-dent was confused and failed to carry out the composition at all. The tenth student reversed the order of elements when carrying out the transformations, and therefore, didn't "have a problem."

Following are some indications of the eight students' attempts to reequilibrate. The first reaction usually was to look for a computational error. When such was not found, the students concluded that the order of applying transformations should have been different from the order of multiplying permutations. Jeff explains:

> Jeff: I think, using the actual square we didn't do them in the right correct order. For some of them it didn't really matter . . .

John explains:

> John: When you do it in permutations, you have to work right to left, but if you're actually going to manipulate the square, you perform the operations left to right.

The idea of "starting on the right" when composing functions may appear strange for novices, and makes sense only with an appropriate interpretation of function composition as $fg(x) = f(g(x))$. A decision to alternate the order, once starting on the left and once on the right, indeed helps with temporary reequilibration. On the other hand it reveals a poor understanding of functions. Thinking of transformations and of permutations as functions would have eliminated the possibility of such a solution.

Resolving disturbing inconsistencies by accepting a "mid-stream order switch" is an approach to which even experienced mathematicians can be reduced. For example, Halmos [7, p. 66] in discussing matrices that correspond to linear transformations acknowledged the "unpleasant phenomenon of indices turning around." Halmos also wrote that "it is a perversity not of the author, but of nature" that makes us use an equation that "works" instead of the "more usual equation." We will return to this example in the following section.

Let us consider briefly the robustness of the students' interpretation. On several occasions, the interviewer pointed out to the student that the rotation R_{90} should have corresponded to [2341], rather than [4123]. In trying to make sense of this remark, the typical response was "Oh, that should have been left/counterclockwise rotation. I thought it was right/clockwise rotation. That's why it didn't work." An excerpt from a student's protocol is given below:

> Int: Let's talk about R_{90}. Why don't you tell me how you came up with [4123] as a permutation.

> Mark: Well R_{90}, we got [4123] because we had a square labeled [1234] starting at the upper left-hand corner. And then we rotated ours clockwise, which gave us, 4 would be in the upper left-hand corner and you'd go [4123] around that way [clockwise]. We rotated ours clockwise. I think you said it should have been, what? [2314]?

[2341], yes. Which would have been a counterclockwise rotation of once around.

The "coincidence" resulting from the fact that R_{90} has order 4 is that the position interpretation of the right turn leads to the same result as the object interpretation of the left turn. When a different correspondence was suggested, the students had a tendency to change the transformation (from right rotation to left rotation) rather than to change the interpretation (from position interpretation to object interpretation). Such behavior suggests that the position interpretation is not just an occasional preference—it may be a very salient interpretation by novices in group theory. If so, then why? Is it an artifact of instruction or is it based in deeper perceptual and conceptual factors? We regard this as a matter deserving close empirically-based attention.

Similar Phenomena

We would like to situate our considerations of two ways to interpret the symmetries of a square with a collection of other phenomena which have some similarities to, but also differences with our situation. We will consider

(1) the alibi/alias dichotomy introduced by Birkhoff and Mac Lane [3, 9] considered in the case of translation in $\mathbf{R}^2$ [3, p. 238] and rotation in $\mathbf{R}^2$ [10, p. 75].

(2) the alibi/alias dichotomy for quadratic forms [3, pp. 250–251; 9, pp. 387–388],

(3) an observation of Halmos on the matrix of a linear transformation [7, p. 65] and

(4) the effect of an automorphism of a vector space on the matrix of a linear transformation.

First, we will describe these phenomena and then compare and contrast them with our object/position dichotomy.

Example 1: Alias/alibi for translation and rotation in $\mathbf{R}^2$.
Birkhoff and Mac Lane [3, p. 238] suggest that an affine transformation of $\mathbf{R}^2$ into itself, can be interpreted as an *alibi*: a transformation, in which each point x is carried into a point y on the same coordinate system, or an *alias*: a change of coordinates, in which the original coordinate system is replaced by a new one. For example, the equations

$$y_1 = x_1 + 2 , \quad y_2 = x_2 - 1$$

can be seen as a translation of every point in the plane two units east and one unit south (alibi) or as a change of the coordinate system to a parallel system with the origin two units west and one unit north of the given origin (alias). We note that both interpretations change the representation, or name, of the point to the same new name: e.g., the point $(0,0)$ is either carried to the point whose coordinates are $(2, -1)$ in the original coordinate system or the point remains in the same place but, in the new coordinate system, it has coordinates $(2, -1)$. We also note that the way this is done in the alibi case is by applying the transformation to points and it is done in the alias case by applying the inverse of the transformation to the coordinate axes (or basis).

A similar example was discussed by Synge [10] who considered the following equations in the Euclidean plane

$$x' = x \cos\theta + y \sin\theta$$

$$y' = y \cos\theta - x \sin\theta$$

as a transformation which can be interpreted in two different ways. "First, we may think of fixed axes, and of a point which at first has coordinates (x, y) and later has coordinates (x', y'), all coordinates being measured with respect to the fixed axes. In this way of looking at the transformation, the axes are fixed and the plane rotates through an angle θ, in the sense from Oy towards Ox. Secondly, we may think of a fixed point, referred first to axes Oxy and secondly to axes $Ox'y'$, the axes $Ox'y'$ being obtained from Oxy by rotating them through an angle θ in the sense from Ox towards Oy. Now the plane is fixed and the axes "rotate"[10, p. 75]. We note here that even though Synge didn't name the interpretations alibi/alias, his two interpretations are essentially the same as those of Birkhoff and Mac Lane. In both cases discussed in Example 1, both the alias and the alibi interpretations lead to the same result in that both change the coordinates to the same new coordinates, and in order to do this, it is necessary to make use of the inverse of the original transformation.

Example 2: Quadratic forms under automorphism.

In a later work, Mac Lane and Birkhoff [9] applied the alibi/alias analysis to several situations such as similar matrices and the signature of a quadratic form. Our next example is adapted from Birkhoff and Mac Lane, [3, pp. 250–251] and Mac Lane and Birkhoff, [9, pp. 387–388]. It discusses what happens to the (symmetric) matrix of a quadratic form with respect to a basis when an automorphism is introduced. Recall that a quadratic form q on a vector space V over $\mathbf{R}$ is a function $q : V \to \mathbf{R}$ such that the following expression defines a bilinear function on $\mathbf{R} \times \mathbf{R}$.

$$q(v + u) - q(v) - q(u), \quad v, u \text{ in } V.$$

Consider a finite dimensional vector space V, a quadratic form q with the symmetric matrix A relative to some basis (x_i). This means that if X is the coordinate function which assigns to a vector v in V the coefficients of its expansion in terms of the basis (x_i), then we have in vector and matrix notation,

$$q(v) = X(v)A(X(v))^T$$

for all v in V. (Here, $X(v)$ is considered to be a "row" vector and $(X(v))^T$ a "column" vector.) Now, suppose we have an automorphism $s : V \to V$, and that its matrix relative to the basis (x_i) is P. This means that $P - (p_{ij})$ and

$$s(x_i) = \sum_j p_{ij}\, x_j.$$

What is the effect on the (symmetric) matrix of the quadratic form v if the automorphism s is applied to V?

We can again make either an alibi interpretation in which we consider that s changes the vectors in v or an alias interpretation in which the vector remains the same, but the coordinates are changed.

For the alibi interpretation, we consider that s changes a vector v to $s(v)$. The coordinates of v are thereby changed from $X(v)$ to the coordinates of $s(v)$, which are $X(s(v))$. Using the transition matrix, we have

$$X(s(v)) = X(v)P$$

Then we can write,

$$
\begin{aligned}
q \circ s(v) &= q(s(v)) \\
&= X(s(v))A(X(s(v)))^T \\
&= (X(v)P)A(X(v)P)^T \\
&= X(v)(PAP^T)(X(v))^T
\end{aligned}
$$

That is, relative to the basis (x_i) the matrix of q is changed to the matrix of $q \circ s$, which is the symmetric matrix PAP^T.

Now for the alias interpretation we may consider that the coordinates of a given vector v are changed from $X(v)$, its coordinates with respect to the basis (x_i), to its coordinates with respect to a new basis (y_i) given by

$$y_i = s(x_i)$$

The new coordinates of a vector v, with respect to the basis (y_i) satisfy

$$X(v) = Y(v)P \quad \text{or} \quad Y(v) = X(v)P^{-1}.$$

So we may write,

$$
\begin{aligned}
q(v) &= X(v)A(X(v))^T \\
&= (Y(v)P)A(Y(v)P)^T \\
&= (Y(v)P)A(P^T(Y(v))^T) \\
&= Y(v)(PAP^T)(Y(v))^T.
\end{aligned}
$$

That is, the matrix of q relative to the basis $(s(x_i))$ is the symmetric matrix PAP^T.

We note that both interpretations result in the same new matrix of the quadratic form. We also note that in going from the alibi to the alias interpretation it is necessary to replace the transformation of coordinates $X(v)P$ to the inverse transformation, that is, $X(v)P^{-1}$.

Example 3: Matrix of a linear transformation.

The following example is discussed in Halmos [7, p. 66]. Consider a finite dimensional vector space V and a basis (x_i) for V. If T is a linear transformation on V, then it can be represented by a matrix. One way to do this is to write out the expansion of each $T(x_i)$ in terms of the basis and lift the coefficients.

Thus, for example if D is the differentiation transformation on the vector space of polynomials of degree $< n$ and we take the monomials for the basis, then we can write

$$
\begin{aligned}
Dx_1 &= 0x_1 + 0x_2 + \cdots + & 0x_{n-1} + 0x_n \\
Dx_2 &= 1x_1 + 0x_2 + \cdots + & 0x_{n-1} + 0x_n \\
Dx_3 &= 0x_1 + 2x_2 + \cdots + & 0x_{n-1} + 0x_n \\
& \quad \vdots \qquad \cdots \qquad & \vdots \\
Dx_n &= 0x_1 + 0x_2 + \cdots + & (n-1)x_{n-1} + 0x_n
\end{aligned}
$$

so that deleting everything but the explicit numbers leads to the matrix,

$$
[D] = \begin{vmatrix}
0 & 0 & 0 & \ldots & 0 & 0 \\
1 & 0 & 0 & \ldots & 0 & 0 \\
0 & 2 & 0 & \ldots & 0 & 0 \\
\cdot & & & & & \cdot \\
\cdot & & & & & \cdot \\
0 & 0 & 0 & \ldots & 0 & 0 \\
0 & 0 & 0 & \ldots & n-1 & 0
\end{vmatrix}
$$

On the other hand, if we wish to use the standard convention of applying a linear transformation to a vector by placing the coefficients of the vector on the right of the matrix as a column vector and multiplying, then we have to first transpose the matrix. Thus, we define the matrix of a transformation T with respect to a basis (x_i) to be (a_{ij}) where i indicates the row, j indicates the column and a_{ij} is the jth component of the expansion of $T(x_i)$ with respect to the basis (x_i). This choice of the transpose of $[D]$ and not $[D]$ to represent the linear transformation was referred to by Halmos [7, p. 66] as "the unpleasant phenomenon of indices turning around."

Example 4: Effect of an automorphism on a matrix of a linear transformation.

Let $T : V \to V$ be a linear transformation and (x_i) a basis for V. Here we will couch our discussion of bases and the representations they provide in the language of duality, rather than transposes and row versus column vectors. We will denote by (a_{ij}) the matrix of T relative to the basis (x_i). It is given by

$$
a_{ij} = < Tx_i, x^*j >
$$

where (x^*j) is the dual basis of (x_i) for the dual V^* of V. Let s be an automorphism of V and we may consider two interpretations of how s might change the matrix. In the first interpretation, s changes the basis (x_i) to the basis (y_i) where $y_i = s(x_i)$. This gives rise to a new matrix (bij) given by

$$b_{ij} = <Ty_i, y_j^*>$$

where (y_j^*) is the dual basis of (y_i) for the dual V^* of V.

In the second interpretation, we consider that s changes T by conjugation: that is, instead of applying T to V, one first applies s, then T and then s^{-1}. In this case, the matrix of T is now the matrix of the conjugated transformation $s^{-1}Ts$ relative to the original basis. Thus we have the new matrix (c_{ij}) given by

$$c_{ij} = <s^{-1}Tsx_i, x_j^*>.$$

Using the fact that the transpose of the inverse of an automorphism is the inverse of the transpose and also the fact that $sy_j^* = x_j^*$, we get:

$$\begin{aligned}
c_{ij} &= <s^{-1}Tsx_i, x_j^*> \\
&= <Tsx_i, s^{-1^*} x_j^*> \\
&= <Ty_i, y_j^*> \\
&= b_{ij}.
\end{aligned}$$

Thus we have an example that is very similar to the alias/alibi situations except that here the two interpretations still give the same new names but it is not necessary to switch to an inverse. One could, of course, argue that there is an inverse present in that transforming the basis (x_i) to the basis $s(x_i)$ amounts to applying the inverse of s to the elements of V. It is not necessary, however, to consider this inverse whereas in the alias/alibi situations, the inverse plays an essential role.

Comparing with Examples 1 and 2, we observe that the use of the inverse transformation in alias situations, to obtain the same new name as in alibi situations, arises only when the objects in the discussion are vectors in a finite-dimensional vector space (points in $\mathbf{R}^2$ in Example 1). This is due to the fact that applying a transformation to a basis results in applying its inverse to the coordinates of elements of a vector space. In Example 2 however, the objects in the discussion are quadratic forms and not vectors. The same automorphism s is applied either to the quadratic form (alibi) and to the basis (alias). The inverse of s appears in the discussion because we chose to perform the calculation at the level of vectors and their coordinates. In the following paragraph, we repeat the analysis of Example 2 avoiding the introduction of the inverse of a transformation by performing the calculations at the level of quadratic forms and their matrix representations.

Revisiting Example 2: Quadratic forms under automorphism.

Let V be a finite dimensional vector space and q a quadratic form on V. Given a basis (x_i) for V, there is a unique symmetric matrix (a_{ij}) which satisfies,

$$q(v) = \sum_{i,j} <v, x_i^*> a_{ij} <v, x_j^*>, \qquad \text{for all } v \text{ in } V.$$

Now suppose that s is an automorphism of V. Then we can interpret the effect of s in two ways. The alias interpretation is that s changes the basis (x_i) to the basis (y_i) where $y_i = s(x_i)$. This gives rise to a new symmetric matrix (b_{ij}) uniquely determined by the relation

$$q(v) = \sum_{i,j} <v, y_i^*> b_{ij} <v, y_j^*>, \qquad \text{for all } v \text{ in } V.$$

In the alibi interpretation, it is considered that s changes q to $q \circ s$. In this case, the matrix of the new quadratic form, relative to the original basis is the unique symmetric matrix (c_{ij}) which satisfies,

$$q \circ s(v) = \sum_{i,j} <v, x_i^*> c_{ij} <v, x_j^*>, \qquad \text{for all } v \text{ in } V.$$

Now, write $w = s(v)$. Using the fact that $s^* y_j^* = x_j^*$ and that the dual of the inverse of an automorphism is the inverse of the dual, we get

$$q(w) = \sum_{i,j} <s^{-1}w, x^*i> c_{ij} <s^{-1}w, x_j^*>$$

$$= \sum_{i,j} <w, s^{-1^*} x_i^*> c_{ij} <w, s^{-1^*} x_j^*>$$

$$= \sum_{i,j} <w, y_i^*> c_{ij} <w, y_j^*>, \qquad \text{for all } w \text{ in } V.$$

By the uniqueness, (b_{ij}) is identical to (c_{ij}). Thus we see that the two interpretations lead to the same new matrix for the quadratic form and in both the alibi and alias interpretations, the automorphism s and not its inverse is used to make the change.

It may be that applying a transformation to points, but the inverse of the transformation to the basis (as in Examples 1 and 2) will appear arbitrary to students and cause some confusion. If that is the case, the fact that in an analysis such as this there is no such switch to the inverse may cause this situation to be less confusing. These are research questions which should be investigated.

Similarities in these situations.

In each of the situations we have described, there are certain objects and in each case there are two naming schemes or means of assigning to an object some algebraic quantity. Thus in Example 1, the objects are the points in $\mathbf{R}^2$ and the first naming scheme is the Cartesian coordinate system. The second naming scheme comes from the introduction of the translation or the rotation. In Example 2 the objects are the quadratic forms and the first naming scheme is the matrix of a quadratic form with respect to a basis. The second naming scheme comes from introducing an automorphism of a vector space. In Example 3 the objects are the linear transformations on a vector space V and the naming schemes are the two ways of lifting the matrix elements from a set of equations—with and without a transpose. In Example 4 the objects are the linear transformations on a vector space V and the first naming scheme is the matrix of a linear transformation with respect to a basis (one of the choices in Example 3). The second naming scheme results from the introduction of an automorphism of a vector space. Finally, in our situation, rigid motions of a square, the objects are the symmetries of the square and the naming schemes are the two ways of assigning a permutation to a symmetry that come from the object/position interpretations.

Before moving over to focus on the differences, we should acknowledge that because of the existence of two naming procedures, there is a potential for mixing them up with a resulting confusion and even error.

Differences among the situations.

One very concrete difference is that in some of these situations such as position/object, and matrix of a linear transformation, the two procedures for naming give different names, whereas in the others, the two new names are the same. In fact, in the first stages of our investigation, after considering Example 1 only as a case of alibi/alias dichotomy, we assumed that the same name was due to the choice to use one transformation in alibi interpretation and its inverse in alias interpretation, which may seem a bit unnatural. If the inverse were not introduced, the two interpretations would give different results. However, our further investigation revealed that the use of the inverse is not essential when moving to a higher level objects, such as quadratic forms or linear transformations.

There is another significant difference, that appears to us more fundamental when comparing the above examples. Consider the object/position situation as opposed to the alibi/alias dichotomy. In the former case, attention is focused on a single object (a symmetry) which is not in any way changed in the discussion. There are, however, two different interpretations that lead to two procedures for naming this object as a permutation. In the latter case, however, attention is again focused on a single object (point on a plane) and a single naming scheme (assignment of Cartesian coordinates), so that no ambiguity exists at first. But then, a transformation is introduced and the question arises of whether to apply it to the objects or to the naming schemes. The two possible answers to this question then give rise to two new procedures for naming the object. Thus, in this second case, one has objects with original names (before the transformation) and then (after the transformation) two possibilities for new names for a total of three names (two of which may be the same) connected with each object. In the former case, there is, from the beginning two names for each object, but that is all that we have. It seems to us that the matrix of a linear transformation is similar in this last respect to the object/position dichotomy while the situation in Example 4 could be another example of the alias/alibi dichotomy.

This last kind of difference seems to us to be essential and we can see no way of embedding all of these types of situations in a single analysis. That identifies one more question to which future research might be directed. However this research comes out, the similarities among the situations appear to be enough to warrant considering them together when searching for effective pedagogical strategies.

We have shown in this paper how difficulties can arise in our situation of object or position interpretations used to assign permutations to symmetries of the square. Synge expresses the opinion that these difficulties do not have to arise in the case of rotations of axes which he considered. He claims that when operations are easily followed intuitively, the change between interpretations is not expected to cause confusion. He makes a point of the importance of making one's interpretation clear in space-time transformations, where "our intuition is not so active" [10, p. 75].

It seems that research is called for regarding the other situations to see if they can be the cause of any student difficulties. If there are such indications, then the question arises as to what pedagogical strategies might help students make sense out of these situations and their multiple interpretations.

Conclusion

Our investigation revealed confusion when Abstract Algebra students attempted to connect symmetries of a square with permutations of S_4. Based on these few students we can say that one possibility is suggested. The position/local interpretation may be more natural for students to use when making a formal correspondence between symmetries and permutations whereas the object/global interpretation may be more natural to use when deciding how to actually move a square (after one or more transformations have been made). Such a situation would, of course, represent an error waiting to be made.

Clearly, transformation groups as well as permutation groups are important examples in the introductory Abstract Algebra course and establishing relationships between the two can be very beneficial for the learner. "The great essential is to try to be quite clear which view we are taking in any particular argument, because otherwise great confusion may result." [10, p. 76]. Therefore, we suggest that the issues discussed above should not be "swept under the rug" or left as exercises, but treated with sufficient attention paid to, and acknowledgment of, the difficulties involved. Of course that is easy to say, but we cannot forget that simply pointing things out to students has very little effect on helping them understand something. Pedagogical strategies must be devised that will help students become aware of these subtleties and use them to make sense out of these important early examples of groups.

However, a more general goal of an Abstract Algebra course is to use the D_n and S_n relationship to introduce mathematics students to broadly applicable skills and invoke or increase their awareness of complexities and non-uniqueness of mathematical interpretations. Our presentation is only an overture to a variety of areas in which such skills or awareness could prove useful. This article is a step in the direction of clarifying a kind of complexity that undergraduate mathematics instructors should be attuned to in order to facilitate the success of their students.

REFERENCES

1. Abelson, H. and diSessa, A., *Turtle Geometry: the Computer as a Medium for Exploring Mathematics*, MIT Press, Cambridge, MA, 1981.
2. Armstrong, M. A., *Groups and Symmetry*, Springer/Verlag, New York, 1988.
3. Birkhoff, G. and Mac Lane, S., *A Survey of Modern Algebra*, 3rd Edition, MacMillan, New York, 1965.
4. Dubinsky, E., *Reflective abstraction in advanced mathematical thinking*, Advanced Mathematical Thinking (D. Tall, ed.), Kluwer Academic Publishers, Boston, MA, 1991.
5. Dubinsky, E., Dautermann, J., Leron, U., and Zazkis, R., *On learning fundamental concepts of group theory*, Educational Studies in Mathematics **27** (1994), 267–305.
6. Fraleigh, J. B., *A First Course in Abstract Algebra*, 4th edition, Addison Wesley, Reading, MA, 1989.
7. Halmos, P. R., *Finite Dimensional Vector Spaces*, D. Van Nostrand Company, Princeton, NJ, 1958.
8. Leron, U. and Zazkis, R., *Of geometry, turtles and groups*, Learning Mathematics and Logo (C. Hoyles and R. Noss, eds.), MIT Press, Cambridge, MA, 1992, pp. 319–352.
9. Mac Lane, S. and Birkhoff, G., *Algebra*, MacMillan, New York, 1967.
10. Synge, J. L., *Relativity: the Special Theory*, 2nd Edition, North-Holland Publishing Company, Amsterdam, 1964.
11. Zazkis, R., Dubinsky, E., and Dautermann J., *Coordinating visual and analytic strategies: A study of students' understanding of the group D_4*, Journal for Research in Mathematics Education (in press).

SIMON FRASER UNIVERSITY

PURDUE UNIVERSITY

CBMS Issues in Mathematics Education
Volume **6**, 1996

To Major or Not Major in Mathematics? Affective Factors in the Choice-of-Major Decision

ANNETTE RICKS LEITZE

Introduction

From the pioneer days in American history to the twentieth century, the three "R"s—reading, 'riting and 'rithmetic—have long been cited as the foundation of American education.

Mathematics is a key that unlocks many opportunities and careers. "Applications, computers, and new discoveries have extended greatly the landscape of mathematics" (National Research Council (NRC) [24, p. 5]).

"Prosperity in today's global economy depends on scientific and technological strength, which in turn is built on the foundation of mathematics education" (NRC [25, p. 1]).

"Advanced mathematics has become more necessary than ever before in old and new fields of science, business, and technology that are vital to America's welfare" (Duren [7, p. 582]). Yet, of the approximately one million high school graduates entering America's colleges and universities with four or more years of secondary mathematics, only about 15,000 major in mathematics [24].

What are some of the factors affecting these students' participation (or lack of participation) in mathematics? There have been numerous studies investigating many variables thought to influence participation in mathematics. Among studies in approximately the last decade, the most popular affective factors have been the following:

(1) peer encouragement or approval (Armstrong and Price [1]; Becker [3, 4]; Lantz and Smith [16]; Lips [17]; Taylor [32]; Treisman [33]),

(2) parental encouragement or approval (Armstrong and Price [1]; Becker [3, 4]; Jayaratne [14]; Lantz and Smith [16]; Lips [17]; Maines, Wallace, and Hardesty [19]; Sherman [27]; Taylor [32]; Yee, Jacobs, and Goldsmith [36]),

(3) teacher encouragement (Armstrong and Price [1]; Becker [3, 4]; Lantz and Smith [16]; Sherman [27]; Taylor [32]),

(4) self confidence (Becker [3, 4]; Hackett [10]; McDade [20]; Sherman [27]; Taylor [32]),

(5) stereotyping of mathematics as male domain (Boswell and Katz [5]; Hackett [10]; Lips [17]; Maines et al. [19]; Sherman [27]; Taylor [32]), and

(6) perceived usefulness of mathematics (Armstrong and Price [1]; Lantz and Smith [16]; Lips [17]; Sherman [27]).

This paper, based on a quantitative and qualitative study of mathematics and nonmathematics majors at a large public research university, discusses some affective factors that may affect participation in mathematics courses leading to a mathematics major. The affective factors include the perceived usefulness, difficulty, asocialness, and enjoyment of mathematics.

Procedures

All subjects in the first (quantitative) component of the study were drawn from the pool of juniors and seniors, enrolled at the University during the 1990-91 school year, who had completed either Calculus I for mathematics/science majors, Calculus II for mathematics/science majors, or both. Within this pool the researcher located two subgroups of students. The first subgroup included those students who had declared mathematics or mathematics education as their major while the second subgroup was composed of a group of nonmathematics, nonscience majors. Nonmathematics, nonscience majors were, for the purposes of this study, defined to be those majors not requiring a 300-level mathematics course. At this university, some examples of courses at the 300 level are Calculus III and IV, Introduction to Differential Equations, and Linear Algebra. Majors excluded from the nonmathematics, nonscience group because they required at least one 300-level mathematics course were mathematics, mathematics education, chemistry, computer science, astronomy, and physics. Mathematics majors were matched to nonmathematics majors on the basis of three factors—gender, calculus professor, and calculus course grade. Such matching eliminates these factors as confounding variables within each pair. Although many potential subjects had to be dropped from the study because a matched pair could not be formed, 70 matched pairs of subjects—140 individuals—resulted from the pairings. Each subject was assigned a code number for anonymity and referential purposes.

Five-point Likert-type surveys were mailed to the 70 matched pairs. Analysis of the 75 returned surveys was accomplished through the use of descriptive statistics. The bulk of the data, however, from the second (qualitative) component of the study, was gathered from a portion of the respondents who agreed to further participation. These 18 respondents were administered individual, hour-long, semi-structured interviews that allowed time for the respondents to feel comfortable in being honest with their answers. All data from the interviews were analyzed using triangulation and a constant comparative method. In triangulation, different researchers examined the data not only to check whether inferences were valid but also to discover which inferences were valid (Hammersley and Atkinson [11]). The constant comparative method— "a process of systematic sifting and comparison" [11, p. 180]—provided the researcher with new questions and conjectures to ask and test in follow-up interviews. Through the constant comparison of data, "eventually, categories and their related properties emerge[d]" (Hutchinson [13, p. 135]).

Affective Factors

Usefulness of mathematics.

With regard to the perceived usefulness of mathematics, there are two observations especially worth noting. The first observation concerns the mathematics and nonmathematics majors' differing group means on the usefulness scale compared with the results of their interviews. The mathematics majors' mean on the usefulness scale of the survey was 1.704 while the nonmathematics majors' mean on the usefulness scale was 1.870. On this five-point Likert scale, a score of 1=strongly agree, 2=agree, 3=undecided, 4=disagree, and 5=strongly disagree. Hence, the lower score of the mathematics majors indicates that they felt more strongly that mathematics is useful than did the nonmathematics majors. Now, that in itself is not noteworthy and, in fact, is to be expected. Among high school students, perceived usefulness of mathematics has been a consistently strong predictor of participation in mathematics (Armstrong and Price [1]; Lantz and Smith [16]; National Institute of Education [23]; Stallings and Robertson [29]).

However, this variable's contribution to college students' decisions has not been thoroughly examined.

Despite the mathematics majors having a lower mean on the usefulness scale than did the nonmathematics majors, in the qualitative portion of the study, not a single mathematics major spoke of how useful mathematics is. The nonmathematics group, however, had much to say about the usefulness of mathematics both in and out of school.

"I've used [math] in every class. I mean I used the math in my stats class. I used it, I use it in my finance classes. I use it constantly." (NMJ123, I-12)

"[Math helped me] to do better on my L[aw] S[cholastic] A[ptitude] T[est], you know." (NMJ103, I-25)

"I think [my mathematics] really helped [me in my other courses] a lot!" (NMS154, II-7)

"You use [math] every day. I think it's important in what I'm gonna be going in to. . . . I don't see how people can think that math cannot be useful cause it can be used in so many different ways." (NMS161, I-16)

"I would not have made it in the Business School or anywhere in the world without being able to do math." (NMJ145, I-15)

The difference between the two groups' beliefs in the usefulness of mathematics seems to be somewhat contradictory and surprising. This difference may be the result of the groups' differing views of the essence of mathematics. Because the nonmathematics majors took little, if any, mathematics beyond calculus, they were faced with little theorem proving and abstract mathematics. For the nonmathematics majors, mathematics is still equation-type, plug-and-chug mathematics—just numbers or arithmetic. It is easy to understand how the nonmathematics majors, with a numbers or arithmetic view of mathematics, were able to see much usefulness of mathematics in numerous aspects of their lives since they encounter numbers or arithmetic on a daily basis. Many of the mathematics majors, on the other hand, had moved beyond the calculus, grind-it-out type of mathematics into more proof-oriented and difficult-to-visualize mathematics where numbers and arithmetic play a less significant role. The mathematics majors' view of mathematics relied heavily on the understanding of mathematical systems and reasoning. Given this view of mathematics, its daily usefulness is much more difficult to envision. Related to the first observation is the second observation dealing with the issue of relevance of mathematics to their careers. The nonmathematics majors, in particular, questioned the relevance issue. The nonmathematics majors largely knew no profession using a mathematics degree.

"I've always heard from teachers and stuff that, you know, you're always gonna need math. . . . And they were always, like, you know, you can do this and it'll probably be easy for you but, you know. . . . Math could be useful and so I should keep going on in calculus rather than go back and take an algebra class in college. . . . [But then I decided] to be done with math." (NMJ128. I-8 and 9)

"I mean I don't think most people have an idea of [what you can do with a math degree]. Other than, I don't know, I, I know some, trying to think of math majors maybe, you know either going into teaching or going into some kind of business or something like that." (NMS134, I-5)

"I think I basically thought that whatever you had your degree in was what you had to do. Like if you were a math major you couldn't be an engineer, you know. Um, I didn't realize then that you could apply it to other places. Like, if you had a math major you could go get a job doing something else or go to school for something else." (NMJ128, II-10)

"I didn't know what you could do with math." (NMJ103, I-25)

Unexpectedly, however, the mathematics majors also were unable to name many professions using mathematics. They could name only two professions using a mathematics degree—actuarial science and teaching. Five of the mathematics majors mentioned knowing about actuarial science and three of the five planned on becoming actuaries. Teaching was also mentioned as a mathematics profession with which the subjects were familiar and two subjects planned to become mathematics teachers. There were no other professions using mathematics mentioned by the group of mathematics majors. One mathematics major explicitly stated, "a lot of the stuff you do in [math], it's just, um, I don't see any use for it" (MS154, II-17). It appears that undergraduates have simply accepted the maxim "mathematics is useful" without seeing evidence of it in their courses. Moreover, usefulness must not be a strong factor in many undergraduates' choice of major if they cannot even name professions using it. Although some people may feel that undergraduates are foolish to spend four or more years of their lives seeking a college degree and not knowing what careers they could pursue, this may, in fact, be the reality of the situation.

It is not clear, however, what, if anything, undergraduates' apparent limited knowledge of possible careers says about the mathematics teaching and/or counseling that the students have received. It may be saying that the mathematics community is falling substantially short when it comes to giving evidence, in their teaching, of mathematics' usefulness. Or it may be saying that high school counselors and/or college advisors are not knowledgeable regarding the usefulness of mathematics. In any case, it is this researcher's opinion that too many students, mathematics majors and nonmathematics majors alike, are unable to give evidence of mathematics' usefulness. Despite this, students continue to verbalize that "mathematics is useful" but mostly ignore its usefulness in choosing a major.

On the basis of these rather provocative findings, this researcher believes that the mathematics majors had extremely limited knowledge regarding avenues open to individuals with mathematics degrees. Chipman and Wilson [6, p. 325] reported that:

> "Almost no one knows what the mathematics studied in high school is really used for. . . . Most students know only that mathematics is required to enter the career they desire, not why it is required or how it will be used."

Results of this study confirm, on the college level, what Chipman and Wilson [6] reported on the high school level. Initially, it appeared to be very curious and very alarming that college mathematics majors did not know what to do with a mathematics degree. Yet, given more time to reflect on the situation, this researcher began to realize that this same situation may, in fact, exist in other disciplines as well. That is, soon-to-be graduates with a major in English, history, a foreign language, or other liberal arts areas may, similarly, not be able to enumerate many professions in which they could be employed. It may be that undergraduates simply want to proceed through their college careers believing that some interesting job will come along after their graduation. Usefulness of mathematics appears not to be a consideration in undergraduates' choice-of-major decision.

Difficulty of mathematics.

Although the difficulty of mathematics was not targeted as a variable this study, this aspect of mathematics continually emerged during the analysis, using triangulation, of the qualitative data. Even though meaning of the word "hard" assumed different forms for different subjects, both the mathematics and nonmathematics majors agreed that mathematics is hard. The subjects' reasons for describing mathematics as hard were because it required a lot of work or was time consuming (MS114, III-10; MS134, II-8; MS151, I-17; NMJ123, I-1; NMJ143, I-5; NMS134, I-1; NMS166, I-19 and 20), was hard to understand (MJ128, II-12; NMS166, I-9), or both (MJ143, II-2; MS154, II-6; NMJ103, I-12; NMJ128, II-14; NMJ145, II-17; NMS154, II-8; NMS161, II-5). One subject expressed this sentiment as follows:

> "It always seemed I walked out of there with eight pages of notes. . . . This is probably the most notes I have taken in my life in a university class. I have taken a lot of notes in other classes, but I went through two notebooks of calculus notes. That stands out."(NMJ145, I-7)

Another subject "felt really overwhelmed with the number of math classes [she] was going to have to take [for a mathematics major]" (NMJ143, I-11).

Some subjects spoke of the reputation of Calculus I for being a hard class (NMJ145, NMJ123).

"My senior year [in high school] I had friends that were here [at the University]. . . . And they took [Calculus I] and they were like, it was a horror story, you know. They were actually, had tremendous amount of trouble with it."(NMJ103, I-1)

In fact, Calculus I has a campus-wide reputation at the University for being a tough class (MS114, II-12). Students begin Calculus I expecting it to be a very difficult course. As MJ143 noted: "In your freshman year [Calculus I] was just supposed to be the hardest class" (I-2).

One explanation for why mathematics is hard, given by a subject in the group of mathematics majors, was that you had to keep up with it. You need a lot of self discipline to do well in mathematics (MJ128, II-2). MS114 expounded on this idea.

"Math is a little different than other subjects. You know a lot of subjects you can be, be—, you can be behind a lot of the year and catch up toward the end. But, but, um math doesn't work like that. You have to, uh, you have to keep up with it every inch of the way."(III-5)

This idea has some merit. The nature of mathematics is that it builds on itself. What is learned today is applied tomorrow. Although it is true that all disciplines are in some ways hierarchical, the hierarchical nature of mathematics seems to be magnified.

The hierarchical nature of other areas, however, is not so magnified. For example, in many introductory courses on British literature, at some point the study of Chaucer's *Canterbury Tales* ends and the study of Shakespeare's *Hamlet* begins. With each new literary work, students have the opportunity to begin afresh. If they lagged behind in reading *Canterbury Tales*, they have the opportunity to start over when reading *Hamlet*. However, in introductory calculus courses, if students lagged behind when studying, say, the derivatives of the standard trigonometric functions (sine, cosine, tangent, cotangent, secant, and cosecant), then each time these derivatives were used in more complicated functions, such as $f(x) = \cos^2 x - \sin^2 x$, the students would be even further confused. Because this type of situation occurs almost daily in most branches of mathematics and at most levels of mathematics, serious problems result when students do not maintain adequate and consistent levels of studying.

The belief that mathematics is hard may have substantial foundation at the University. Figures compiled by the University's Office of the Registrar list the mean grade point average (GPA), by semester, given to undergraduates enrolled in undergraduate courses in the College of Arts and Sciences' academic departments. Grades assigned in the Department of Mathematics were consistently among the lowest of all lower division courses (100- and 200-level) in the 41-46 departments listed in these reports. Of the ten semester reports examined for this study (1986-1991), the GPA in lower division mathematics courses was in the lowest five percent for six semesters. Of the other four semesters, the GPA in lower division mathematics courses was consistently in the lowest ten to fifteen percent. Even though no respondents mentioned a relationship between mathematics being hard and receiving low grades, such a relationship may exist. Assuming this relationship, it appears that undergraduates' beliefs regarding lower division mathematics courses at the University may be well founded.

Other studies, but not necessarily studies on mathematics participation, have found similar results. Burton's study (cited in [26]) supports the view of mathematics as hard, mentioned by many of the subjects in this study. Furthermore, Barnes and Coupland's study [2] identified two problems associated with introductory calculus courses that caused those courses to be labeled as difficult: students' lack of motivation to study mathematics and conceptual difficulties in the introductory parts of the course. Barnes and Coupland [2] attributed students' lack of motivation to the insufficient attention given to real-world applications in the initial stages of introductory calculus texts and attributed the conceptual difficulties to the introduction of the idea of a limit. In their view

> "The definition of a limit requires considerable mathematical sophistication before it can be properly understood, so a formal treatment of limits is not appropriate for most beginning students. . . . When students finally learn that differentiation can be reduced to a fairly simple list of rules which can be applied mechanically, they are usually very relieved and often decide to ignore everything that went before, and simply memorize the rules. . . . The result of the usual approach to beginning calculus, therefore, is that many students give up trying to understand, and resort to instrumental learning, their only purpose being to get through an examination and then forget it all." (p. 73)

Thus, introductory calculus, taught using traditional methods and texts, promotes the idea that mathematics is hard.

But for NMS166 hard work on a hard subject was not enough to ensure success. Success in mathematics also required innate ability (NMS166, I-19). And other subjects agreed that mathematics ability was innate (viz., MJ143, I-16; MS114, III-4; MS151, II-20). This belief in innate ability, while common among American students and parents (NRC [25], Stevenson and Lee [31]), is not common among Asian students and parents. In Asian philosophy, "differences in innate ability are deemphasized, and the potential for change throughout life is believed to lie within the individual" (Stevenson, [30, p. 8]). Hence, Asian children learn at an early age that they must work hard to achieve success [30, 31].

Perhaps the undergraduates' belief that they had a gift for math explains why some subjects showed perseverance by continuing in mathematics in the face of discouragement. Four different subjects failed one or more mathematics classes in their college career. Two of the four, MJ143 and MS114, believed they had a gift for math. The third, MJ145, expressed considerable confidence in mathematics but indicated no adherence to the belief that he had mathematics ability while the fourth, MS154, just gave up working in the class in which he was not doing well but did not give up mathematics overall. MS114 even suggested the reason for the relatively low number of mathematics majors was that people don't have the perseverance for it.

In spite of the unanimous opinion of the subjects in the qualitative portion of the study that mathematics is hard, only one alluded to this as a factor in her choice-of-major decision. That subject was overwhelmed by the amount of mathematics courses that would have been required of her had she selected mathematics as a major. No mention of the aspect of mathematics being difficult was made by any other subjects in their choice-of-major decision.

Enjoyment of field of study.

Questions designed to probe the area of choice of major generated enjoyment of the field as the most prominent reason for pursuing a particular field of study. Both the mathematics and nonmathematics majors most often cited enjoyment of high school and/or introductory college classes in a given discipline as the reason for selecting that field of study as a major (MJ128, I-8; MJ143, I-10; MS134, I-5 and 6; MS151, I-16; MS154, I-3; NMJ103, I-10; NMJ123, I-6; NMJ128, I-8; NMS134, I-4 and 5; NMS151, II-4; NMS154, I-9; NMS161, I-1; NMS166, I-6 and 7). Several subjects elaborated on their decision to choose a major based on their enjoyment of the field.

> "Math was my, my favorite class, what I really enjoyed more and nothing else stood out. I mean it was like head and shoulders above the rest of anything that I really wanted to go in to. "(MS134, I-6)

> "I don't enjoy school very much but math was one of the classes that I guess I enjoyed the most."(MJ143, I-10)

> "When I was taking [Calculus II] I took two education classes. And I really liked them. They were pretty interesting. They were a lot of fun . . . so I just kept it up."(MS151, I-16)

> "I really hated the [computer science] professor. I really hated the class. Meanwhile I liked my math class."(MJ128, I-8)

The above quote by subject MJ128 is related to another subject's (NMS151) astute observation regarding the use of enjoyment as a gauge by which to choose a major. He noted that any field of study should be considered separate from the professor (NMS151, II-5). That is, to make a decision for or against a given discipline as a major purely on the basis of enjoyment of a professor is not wise. Although NMS151 is now able to differentiate between enjoyment of a field of study and enjoyment of a professor, he indicated that he was not able to make that distinction at the time he was choosing a major.

Evidently, other students, at the time they were faced with the choice-of-major decision, also were unable to differentiate between enjoying the professor and enjoying the field of study. Other subjects in the study spoke of transferring either positive or negative feelings for a professor onto a field of study.

> "My genetics professor, he didn't ever seem like he wanted to be in there and, uh, he always blamed it on the class. But, uh, I never really wanted to get into genetics after that." (NMS154, I-13)

> "You know to get you interested or to get you, you know, I think [having a good professor in your introductory level classes] makes all the difference." (NMS134, I-23)

Piquing individuals' interest in or enjoyment of a given discipline is a necessary but not sufficient condition to ensure that they will remain interested. Of the subjects in the qualitative portion of the study, 15 out of 18, or 83.3 percent, changed majors one or more times during their college careers. It seems to this researcher, that this is a staggering number of indecisive undergraduates.

If, in fact, undergraduates are transferring positive or negative feelings about the professor onto the field of study, and subsequently using enjoyment as a basis for selecting a major, that might help explain the widespread practice of changing majors one or more times. Consider the following scenario. An undergraduate takes an introductory course in department B from professor B_1. After studying under professor B_1 for one or two semesters (a typical length of introductory course work), the undergraduate enjoys the professor, transfers those feelings of enjoyment onto the class, and hence, decides to major in discipline B. Then after studying under professor B_2 or B_3, one or both of whom the undergraduate does not particularly enjoy, (s)he transfers the negative feelings for professor B_2 or B_3 onto discipline B, and decides once again to change majors—getting out of discipline B. This type of scenario certainly explains the frequent (almost semester-by-semester) change of major some undergraduates experience. Even though results of this study support the enjoyment of discipline as the primary basis for choosing a major, the transfer of feelings explanation for undergraduates' frequent change of major is, at this point, purely hypothetical. However, given the impact that enjoyment plays in undergraduates' choice of major, it is a scenario worthy of further investigation.

Although Rodgers and Mahon's study (cited in [26]) found enjoyment of mathematics to be the most common reason for choosing to study Advanced (or A-level) mathematics in Ireland, the remaining literature was conspicuously void of this aspect of choice of major in American colleges and universities. In the United States, it appears that one of two cases may have resulted. First, it may be the case that researchers have simply ignored the enjoyment factor in the choice-of-major decision. Or second, since enjoyment is difficult to uniformly describe and measure, it is possible that researchers have investigated the enjoyment factor by examining other closely related factors such as self confidence, which may contribute to one's feelings of enjoyment. In either case, enjoyment turned out to be the single most salient factor in the choice-of-major decision and, thus, may be a topic deserving further research.

Mathematics is asocial.

Although the researcher originally had no intent to investigate the asocialness of mathematics as a topic of study, this view of mathematics emerged in the triangulation of the qualitative data. Since positive attitudes toward mathematics have been found to increase the likelihood of participation in high school and/or college level mathematics (Armstrong and Price [1]; Becker [3, 4]; Ethington and Wolfle [8]; Sherman [28]), subjects' views of mathematics as asocial are deserving of investigation and discussion. Although both the mathematics and nonmathematics groups commented on certain aspects of their experiences that demonstrated their feeling of mathematics as asocial, the nonmathematics group adhered to this belief more strongly than did the mathematics group.

The unanimous opinion of the nonmathematics majors was that mathematics is an asocial activity.

> "[Math] seemed like too lonely of a job. . . . And math has always been plug and chug on your own." (NMJ103, I-26)

> "Sitting there and reading the problem by myself." (NMJ128, I-7)

> "You have to do a lot of the work [in college mathematics] on your own." (NMS161, II-2)

Group work was not encouraged nor carried out in subjects' introductory level college mathematics courses (NMJ103, I-17; NMJ128, II-5; NMJ143, I-9; NMJ145, I-15; NMS134, I-19; NMS154, I-21). Only one subject had a professor who encouraged group work and that was in a 300-level course, not an introductory course (NMJ143, I-19). Their introductory mathematics courses consisted of attending lecture, attending recitation, and doing homework "despite mounting evidence that the lecture-recitation method works well only for a relatively small proportion of students" (NRC [25, p. 17]). This one dimensional approach to learning does not easily accommodate diverse learning styles and perpetuates the notion of mathematics as asocial.

Some subjects' views of mathematics were further evidenced in their asocial descriptions of mathematicians.

> "[Mathematicians] just to sit around and [work on] why equations work this way and what laws you can derive from a bunch of different facts put together." (NMJ103, I-20)

"Someone that is, works on math problems all day long. . . . Kind of like in a laboratory. I think of it as a scientist. Studying. White robe type thing. . . . Just a lot of numbers. . . . Just works on it all the time. . . . I don't think he has a family." (NMJ128, II-12)

"[Mathematicians are] sort of quiet, sort of to themselves. Um, like if I would see a mathematician in the library or whatever I would picture them with their nose in the book, or with big sheets of paper working on problems." (NMJ145, II-2)

While most of the mathematics majors did not specifically admit to an asocial view of mathematics, one can infer certain attitudes and beliefs regarding their feelings about mathematics by examining their experiences in mathematics and their view of mathematicians. Many of the mathematics majors pictured a mathematician as a male doing a lot of solitary work. Two mathematics majors responded that mathematicians were strange people but were careful to add that they themselves were not strange like the mathematicians. One mathematics major even went so far as to say that mathematicians have no concept of the world—"they stay in every night and they do this stuff" (MJ128, I-14). So even though most of the mathematics majors did not explicitly talk about mathematics as being asocial, their attitudes and experiences promote the idea of mathematics as a solo or asocial activity!

Actually the mathematics majors' descriptions of how they learned mathematics strengthened the nonmathematics majors' asocial view of mathematics. Both groups described their introductory mathematics courses not only as lacking professor-student interaction but also as void of group work. Although mathematics professors don't really discourage group work, they don't really encourage it either (MJ103, II-5). "[Mathematics professors] don't really talk much about how you're supposed to get math" (NMS166, I-14). Despite the increased use of collaborative assignments in many college mathematics classrooms, the teaching/learning model displayed in the classrooms in which most of the subjects of this study sat indicated to them that learning mathematics was an individual effort. Only one subject in the study, a mathematics major, preferred to work and study alone rather than in a group. This student, an athlete, felt too much time was wasted in group work and she simply preferred to go talk to the professor (MS134, I-9). The remainder of the subjects, both mathematics and nonmathematics majors in the qualitative portion of the study, participated or preferred to have participated in group study. The few subjects that reported doing some group work in mathematics courses applauded its benefits; the others lamented its absence.

A cursory comparison of school level courses reveals similar results. A typical secondary English class may, indeed, consist of individual reading of some literary work. However, such reading is generally followed by whole class or small group discussions of the reading. In the average secondary history or social science classroom, discussions or debates are likely to occur. Although the science classroom may consist largely of lecture format, it is invariably accompanied by lab experiments frequently carried out by teams of students. In all of these cases, interaction among participants and exchange of ideas take place. In the mathematics classroom, however, the typical scene is of lecture format, followed by individual work on problems.

The college classroom experiences of the subjects in this study served to affirm their previous experiences. While working together may be a fairly common occurrence in courses in other departments such as music (NMS166), education (MS151), philosophy, and history (MJ128), the college mathematics classroom is still largely an individual effort. One subject pursuing a double major, mathematics and music, summed up her comments this way.

> "It seemed like in a lot of my math classes that most people wanta go in there and take notes and you know, and leave. Which is so much different than from say my music courses where you live with these people. And I have never been in a music course where people didn't get to know one another and spend a lot of time together, study together, helped each other through everything. But I haven't found that in math classes. . . . Some people I've tried to get to know but, you know, well [they'd say] I'll do my studying and you do yours." (NMS166, I-10)

The difference between the nonmathematics majors' frank perceptions of mathematics and mathematicians as asocial, and the absence of same from the mathematics majors may be related to each group's differing view of what it means to do mathematics. Recall that the nonmathematics majors felt that to do mathematics means to use set procedures to produce numeric answers (i.e., arithmetic)—an activity transpiring between oneself and a pencil, with little or no need for human-to-human interaction. It is easy to understand how individuals who view mathematics as just numeric-producing procedures might also view mathematics as asocial. The mathematics majors' view of mathematics, however, which relied on understanding mathematical systems and reasoning, is consistent with a social view of mathematics. The absence of an articulated asocial view of mathematics by the mathematics majors, in spite of the solitary way that they learned mathematics, may be representative of conflict they feel between their view of mathematics and their own personal experiences. That is, perhaps they feel the need to communicate and interact to gain mathematical understanding and reasoning but their classes continue to model the lecture format.

Representing a view of learning much different from that practiced by many college mathematics instructors is L.S. Vygotsky and his followers. According to John-Steiner and Souberman [15], many of Vygotsky's writings, only some of which are available in English, "further his fundamental hypothesis that the higher mental functions are socially formed and culturally transmitted" [15, p. 126].

Vygotsky and his followers wrote much about collaborative learning and what he called the zone of proximal development, which is explained in the following excerpt.

> "We propose that an essential feature of learning is that it creates the zone of proximal development; that is, learning awakens a variety of internal developmental processes that are able to operate only when the child is interacting with people in his environment and in cooperation with his peers. Once these processes are internalized, they become part of the child's independent developmental achievement." [35, p. 90]

But what does Vygotsky's work, which dealt mostly with preschool and elementary aged children, have to do with teaching undergraduates? Gallimore and Tharp [9, p. 186] contend

> "The lifelong learning of any individual is made up of these same regulated, [zone of proximal development] sequences—from other-assistance to self-assistance—recurring over and over again for the development of new capacities. For every individual, at any point in time, there will be a mix of other-regulation, self-regulation, and automatized processes."

Hence, some researchers believe that peer collaboration and the zone of proximal development applies to teaching people of all ages and not just young children. The classroom implication of these hypotheses is that the use of collaborative learning situations is an effective means of furthering students' development (Gallimore and Tharp [9], Luria and Vygotsky [18], NCTM [21, 22]; Tudge [34]; Vygotsky [35]). Even though Vygotsky and his followers' loyalty to socially formed learning was not based on a desire to promote positive attitudes toward learning, positive attitudes may well be an indirect benefit.

Perhaps the real issue here is not one of whether mathematics teaching should consist only of collaborative methods, or only of lecture methods, but rather an issue of teaching mathematics using some combination of the two methods. Certainly, strong arguments could be made for continuing to teach mathematics in the traditional way. But equally strong arguments might be made against the traditional way. The strongest point to be made here is that by incorporating collaborative learning into the classroom, diverse learning styles are accommodated and more positive attitudes about mathematics are promoted.

Many professionals are working to incorporate collaborative learning into mathematics classrooms as part of the current reform movement. At the K-12 level, the NCTM's *Curriculum and Evaluation Standards for School Mathematics* [21] and *Professional Standards for Teaching Mathematics* [22] address the need for interactive instructional approaches affording students the opportunity to communicate their mathematical ideas to the teacher and to one another. Many of the NCTM's ideas may apply to undergraduate mathematics education as well. At the undergraduate level, mathematics departments and faculty have been charged to "explore effective alternatives to 'lecture and listen' . . . , involve students actively in the learning process . . . , and to employ varied instructional approaches: group methods, writing, investigative assignments, laboratory projects" (NRC [25, p. 34]).

As long as mathematics is viewed as the exception, rather than the rule, when it comes to group versus solitary activities, mathematics will also likely be viewed as asocial. According to Elkonin (cited in [12, p. 357]), "the late school and youth period [including the first years of college] is characterized by the development of motives for social and societal involvement and methods for mastery of personal relations as well as work and societal requirements." If this is, in fact, a characteristic of undergraduates and if enjoyment is a strong factor in undergraduates' choice-of-major decision, participation in a discipline viewed as a solitary endeavor is likely to be limited.

Summary and Conclusions

While all respondents in the study repeatedly spoke of the usefulness of mathematics, in the qualitative portion of the study, the nonmathematics majors adhered more strongly to the belief that mathematics is useful than did the mathematics majors. In the quantitative portion of the study, the group of mathematics majors had a lower mean (indicating a stronger belief that mathematics is useful) on the usefulness scale than did the nonmathematics majors. Certainly one aspect of the usefulness of any discipline is being able to name professions using that discipline. Yet, the mathematics and nonmathematics majors could name at most two professions using mathematics. The statement "mathematics is useful" appeared to be more of an automated response rather than a belief shaped by the undergraduates' college mathematics experiences.

In spite of the undergraduates' firm conviction about how useful mathematics is, not one subject in the study mentioned a discipline's perceived usefulness as being important in the choice-of-major selection process. This is in contrast to much of the literature which suggests that mathematics' usefulness is important to make known to students. In view of these contrasting results, this may be a very important area for further research.

Both the mathematics and nonmathematics majors cited enjoyment of courses as the most salient factor in selecting a major. This does, in fact, seem like a reasonable way of choosing one's vocation. However, enjoyment is such an elusive entity that it makes measuring enjoyment a very difficult task. The literature in mathematics education reflects this difficulty since it mostly ignores this variable's influence. Mathematicians and mathematics educators investigating participation in mathematics may want to find ways to examine and measure the elusive factor we call enjoyment.

Mathematics was viewed by many subjects in this study as asocial—a solitary endeavor. This is not a particularly endearing descriptor. Humans are generally a gregarious species—they desire companionship and enjoy interaction with other people. The undergraduates in this study similarly enjoyed courses where interaction transpired. But their college mathematics classrooms were primarily of lecture format with little or no human-to-human interaction. It is no wonder that mathematics (and consequently mathematicians, since they love to do mathematics) was viewed as asocial.

These findings combine to provide a further critical result: Introductory level courses (i.e., lower division courses) are vitally influential in determining undergraduates' choice of major. Unfortunately, however, many of these courses are precisely the ones that departments of mathematics refer to as service courses and are generally thought of as "less than desirable" to teach. Such a faculty attitude may be contributing to the notion of mathematics as a filter rather than as a pump. In fact, one mathematics major spoke his feelings by frankly saying,

> "[Mathematicians] aren't interested in making more and more and more mathematicians. Although that, even if our society needs more they aren't necessarily interested in making more. Neither are scientists or anyone else. So, I think it's, it's kind of a social Darwinistic approach. They, survival of the fittest, as far as that is concerned." (MS114, III-6)

Individuals concerned with promoting the notion of mathematics as a pump in our nation's colleges and universities may want to examine areas outside of collegiate mathematics. Looking to the pedagogy employed in classrooms outside of the scientific disciplines may provide those people interested in undergraduate mathematics education with some fresh ideas on how to improve their own instruction and how to promote mathematics as a pump rather than as a filter.

References

1. Armstrong, J. M. and Price, R. A., *Correlates and predictors of women's mathematics participation*, Journal for Research in Mathematics Education **13** (1982), 99–109.

2. Barnes, M.; Coupland, M., *Humanizing calculus: A case study in curriculum development*, Gender and Mathematics (L. Burton, ed.), Cassell, London, 1990, pp. 72–80.

3. Becker, J. R., *The pursuit of graduate education in mathematics: Factors that influence women and men*, Journal of Educational Equity and Leadership **4(1)** (1984), 39–53.

4. Becker, J. R., *Graduate education in the mathematical sciences: Factors influencing women and men*, Gender and Mathematics (L. Burton, ed.), Cassell, London, 1990, pp. 119–130.

5. Boswell, S. and Katz, P., *Nice girls don't study mathematics: Final report*, Boulder, CO: Institute for Research on Social Problems (ERIC Document Reproduction Service No. ED 188 888) (1980).

6. Chipman, S. F. and Wilson, D. M., *Understanding mathematics course enrollment and mathematics achievement: A synthesis of the research.*, Women and mathematics: Balancing the Equation (S. F. Chipman, L. R. Brush, and D. M. Wilson, eds.), Lawrence Erlbaum Associates, Hillsdale, NJ, 1985, pp. 275–328.

7. Duren, W. L., Jr., *The most urgent problem for the mathematics profession*, Notices of the American Mathematical Society **41** (1994), 582–586.

8. Ethington, C. A. and Wolfle, L. M., *Women's selection of quantitative fields of undergraduate study: Direct and indirect influences*, American Educational Research Journal **25(2)** (1988), 157–175.

9. Gallimore, R. and Tharp, R., *Teaching mind in society: Teaching, schooling, and literate discourse.*, Vygotsky and Education, Instructional Implications and Applications of Sociohistorical Psychology (Luis C. Moll, ed.), Cambridge University Press, Cambridge, England, 1990, pp. 175–205.

10. Hackett, G., *Role of mathematics self-efficacy in the choice of math-related majors of college women and men: A path analysis*, Journal of Counseling Psychology **32(1)** (1985), 47–56.

11. Hammersley, M. and Atkinson, P., *Ethnography: Principles in Practice*, Routledge, London, 1989.

12. Hedegaard, M., *The zone of proximal development as basis for instruction*, Vygotsky and Education, Instructional Implications and Applications of Sociohistorical Psychology (Luis C. Moll, ed.), Cambridge University Press, Cambridge, England, 1990, pp. 349–371.

13. Hutchinson, S. A., *Education and grounded theory*, Qualitative Research in Education: Focus and Methods (R. R. Sherman and R. B. Webb, eds.), Falmer Press, London, 1988, pp. 123–140.

14. Jayaratne, T. E., *The impact of mother's math experiences on their daughters' attitudes toward math*, Paper presented at the Biennial Meeting of the Society for Research in Child Development, April 1987, Baltimore, MD.

15. John-Steiner, V. and Souberman, E., *Afterword*, Mind in society: The Development of Higher Psychological Processes (M. Cole, V. John-Steiner, S. Scribner, and E, Souberman, eds.), Harvard University Press, Cambridge, MA, 1978, pp. 121–133.

16. Lantz, A. E. and Smith, G. P., *Factors influencing the choice of nonrequired mathematics courses*, Journal of Educational Psychology **73** (1981), 825–837.

17. Lips, H., *The role of gender, self- and task perceptions in mathematics and science participation among college students* (1988), Social Sciences and Humanities Research Council of Canada (ERIC Document Reproduction Service No. ED 297 945), Ottawa, Ontario.

18. Luria, A. R. and Vygotsky, L. S., *Ape, Primitive Man, and Child: Essays in the History of Behavior*, (translated by E. Rossiter, original work published 1930), Paul M. Deutsch Press, New York, 1992.

19. Maines, D. R., Wallace, J. J., and Hardesty, M., *Attrition from mathematics as a social process*, (Report No. NIE-G-81-0029) (1983), Northwestern University, Program on Women (ERIC Document Reproduction Service No. ED 237 342), Evanston, IL.

20. McDade, L. A., *Knowing the "right stuff": Attrition, gender and scientific literacy*, Anthropology and Education Quarterly **19** (1988), 93–114.

21. National Council of Teachers of Mathematics, *Curriculum and Evaluation Standards for School Mathematics*, Reston, VA, 1989.

22. National Council of Teachers of Mathematics, *Professional Standards for Teaching Mathematics*, Reston, VA, 1991.

23. National Institute of Education, *A national assessment of achievement and participation of women in mathematics: Final report*, (Report No. NAEP-10-MA-60) (1979), Education Commission of the States (ERIC Documentation Reproduction Service No. ED 187 562), Denver, CO.

24. National Research Council *Everybody counts: A report to the nation on the future of mathematics education*, National Academy Press, Washington, DC, 1989.

25. National Research Council, *Moving beyond myths: Revitalizing undergraduate mathematics*, National Academy Press, Washington, DC, 1991.

26. Rodgers, M., *Mathematics: Pleasure or pain?*, Gender and Mathematics (L. Burton, ed.), Cassell, London, 1990, pp. 29–37.

27. Sherman, J., *Factors predicting girls' and boys' enrollment in college preparatory mathematics*, Psychology of Women Quarterly **7** (1983), 272–281.

28. Sherman, J., *Girls talk about mathematics and their future: A partial replication*, Psychology of Women Quarterly **7** (1983), 338–342.

29. Stallings, J. and Robertson, A., *Factors influencing women's decisions to enroll in advanced mathematics courses: Final report*, (Report No. SPI-7009) (1979), SPI International (ERIC Document Reproduction Service No. ED 197 972), Menlo Park, CA.

30. Stevenson, H. W., *America's math problems*, Educational Leadership **45(2)** (1987), 4–10.

31. Stevenson, H. W. and Lee, S. Y., *Contexts of achievement*, Monographs of the Society for Research in Child Development **55** (1990), (1-2, Serial No. 221).

32. Taylor, L., *American female and male university professors' mathematical attitudes and life histories*, Gender and Mathematics (L. Burton, ed.), Cassell, London, 1990, pp. 47–59.

33. Treisman, P. U., *Teaching mathematics to a changing population: The professional development program at the University of California, Berkeley (Part 1: A study of the mathematics performance of black students at the University of California, Berkeley)*, Mathematicians and Education Reform: Proceedings of the July 6-8, 1988 Workshop (N. Fisher, H. Keynes, and P. Wagreich, eds.), American Mathematical Society, Providence, RI, 1990, pp. 33–46.

34. Tudge, J., *Vygotsky, the zone of proximal development, and peer collaboration: Implications for classroom practice*, Vygotsky and education, Instructional Implications and Applications of Sociohistorical Psychology (Luis C. Moll, ed.), Cambridge University Press, Cambridge, England, 1990, pp. 155–172.

35. Vygotsky, L. S., *Mind in Society: The Development of Higher Psychological Processes*, (M. Cole, V. John-Steiner, S. Scribner, and E. Souberman, Eds. and Trans., original works published 1930-1966), Harvard University Press, Cambridge, MA, 1978.

36. Yee, D. K., Jacobs, J. and Goldsmith, R., *Sex equity in the home: Parents' influence on their children's attitudes about math*, Paper presented at the Annual Meeting of the American Educational Research Association, San Francisco, CA ((April 1986)).

DEPARTMENT OF MATHEMATICAL SCIENCES, BALL STATE UNIV., MUNCIE, INDIANA

CBMS Issues in Mathematics Education
Volume **6**, 1996

Success in Mathematics: Increasing Talent and Gender Diversity Among College Majors

MARCIA C. LINN AND CATHY KESSEL

ABSTRACT. Over half the students who select mathematics as a college major switch to other fields. Although equal numbers of males and females enter college intending to be mathematics majors, females comprise 43% of those completing the undergraduate degree and 20% of those completing the Ph.D. We synthesize findings from two lines of research to shed light on this pattern of participation and persistence in mathematics. First, we examine grades earned by males and females as well as by persisters and switchers in mathematics. Studies of over 39,000 students indicate that females earn higher mathematics grades than males in undergraduate mathematics courses. In addition, switchers earn grades equal to those of persisters and some of the most talented males and females switch out of mathematics.

Second, we examine interview studies of over 1,500 students interested in mathematics and science from over 20 colleges and universities. In every study undergraduates complain that mathematics courses are designed to "weed out" students rather than to encourage the best to persist. Switchers more than persisters point out the poor quality of undergraduate mathematics instruction compared to instruction in other courses. Quality of instruction more than success in mathematics motivates students to switch out of mathematics and more females than males are so motivated. A consequence of the perceived low quality of mathematics instruction is the loss of talented students to other majors. We discuss options for calculus reform as a remedy for loss of talent and decreased diversity among mathematics majors.

Thanks to Madeleine Bocaya, Anna Chang, Dawn Davidson, Tiffany Davis, Patricia Kim, and Jean Near for assistance with the production of this manuscript. An earlier version of this work was presented at the 1995 AERA Annual Meeting, Division D Symposium, ("Gender and Mathematics Performance," Organizer, Katherine Ryan), Friday, April 21, in San Francisco, California; and also at the 1995 American Mathematical Society Conference, at the Hebrew University, Wednesday, May 24, in Jerusalem, Israel. This material is based upon research supported by the National Science Foundation under grants MDR-8954753 and MDR-9155744. Any opinions, findings, and conclusions or recommendations expressed in this publication are those of the authors and do not necessarily reflect the views of the National Science Foundation.

Introduction

Two-thirds of the students who plan to study mathematics in college eventually choose other fields. Furthermore, gender but not grades or test scores predicts who switches out of mathematics. Based on their undergraduate grades in mathematics, women should predominate in mathematics careers, but the reverse is true. Women compared to men earn higher grades in college where they comprise 43% of the mathematics majors, yet only 3% of the faculty at the "top ten" mathematics departments are women. Of the students entering graduate school in mathematics, 35% are women, yet only about 20% of Ph.D.'s in mathematics go to females. Why do talented and diverse students switch out of mathematics? To address this question we synthesize results from research on mathematics course experience and mathematics course reform.

In particular, to improve understanding of persistence and diversity in college mathematics we look at mathematics teaching and learning. We examine undergraduate mathematics grades in courses. We synthesize interview studies to explain persistence among undergraduates, and we analyze differences among institutions in retaining mathematics majors. We then analyze recent reforms of mathematics instruction, identifying promising improvements. We also assess college entrance examinations in mathematics, seeking an explanation for the lower scores earned by females in contrast to their grades. Finally, we consider the context of mathematics learning, analyzing how stereotypes, beliefs, and expectations distinguish the experiences of males and females, and suggesting directions for course improvement.

We draw on recent research on learning and instruction as we review this body of research. We weigh the conceptual and cultural experiences of mathematics students. We analyze the processes students follow in making sense of mathematical ideas. We reflect on how students respond to mathematics learning experiences. We pay particular attention to the views that students develop about their own mathematics learning and about their place in the mathematics community.

Undergraduate grades and persistence of males and females

A synthesis of studies at colleges and universities across the United States demonstrates that female students have higher undergraduate grade point averages than male students, that institutions vary in their ability to retain females, and that males more than females continue to study mathematics. Female superiority in course grades begins in high school [57, 69] and holds for mathematics, science, and technology courses as well as courses in the humanities [7, 50, 64, 131].

Most faculty would agree that mathematics grades are an excellent indicator of promise for an undergraduate student in mathematics. Grades are awarded in undergraduate courses by knowledgeable members of the mathematics profession who have the authority to set standards for students. Women are meeting these standards proportionally more frequently than men. Yet women are less successful in entering and completing graduate programs than men. We look closely at grades in this section and at impressions of undergraduate courses in the next section.

Grades earned.

Beginning in high school and continuing through college, females earn higher grades in mathematics, science, and technology courses as well as courses in the humanities. To substantiate this point we synthesize grade data from a variety of research [7, 50, 57, 63, 68, 131].

Studies show that females earn higher grades in general and in mathematics in particular. Significant overall differences in college GPA between males and females are reported in a broad range of studies as shown in Table 1 and Figure 1.

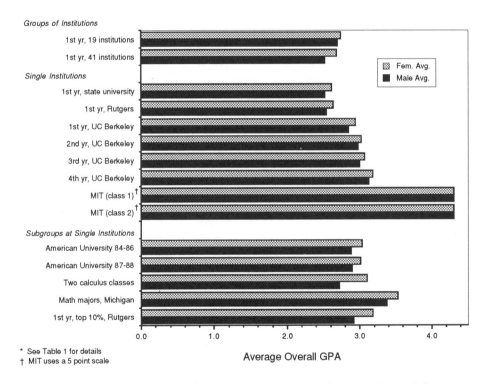

FIGURE 1. Overall GPAs earned by undergraduate males and females at various institutions.

In addition, a diverse set of studies show that females perform as well as or better than males in required and advanced mathematics courses (see Table 2 and Figure 2). This pattern holds for courses in college algebra, precalculus, calculus, courses beyond calculus, as shown in Table 2 [5, 33, 65, 138]. In addition, studies of mathematics majors and students in the top 10% of the university report that

TABLE 1. Overall GPAs earned by undergraduate
males and females at various institutions.

	year	N	% Female	Overall College GPA Males	Females
Groups of Institutions					
1st year students at 19 colleges and universities [119, p. 8]	1980	5,897	44	2.69	2.74
1st year students at 41 colleges and universities [17, p. 21]	1980	68,142	51	2.52	2.68
Single Institutions					
1st semester students at large state university [119, p. 8]	1988	4,307	53	2.52	2.61
1st year students at Rutgers [52, p. 5]	1985	1,956	51	2.54	2.64*
1st year students at UC Berkeley	1992	3,404	49	2.85	2.94
2nd year students at UC Berkeley	1992	4,243	46	2.97	3.02
3rd year students at UC Berkeley	1992	6,079	48	3.00	3.06
4th year students at UC Berkeley [92, pp. 43, 50]	1992	7,646	46	3.12	3.18
MIT students who completed degree and entered in same year (class 1)		1,000 -1,050	27	4.30	4.30
MIT students who completed degree and entered in same year (class 2) [50, p. 75]		1,000 -1,050	38	4.30	4.30
Subgroups at Single Institutions					
Undergraduates at American University who took arithmetic or algebra test	1984 -1986	1,692	61	2.88	3.03*
Undergraduates at American University who took arithmetic or algebra test [109, pp. 79, 84]	1987 -1988	2,937	61	2.89	3.01*
Students in two calculus classes, University of Colorado, Boulder [120]	1973	154	27	2.72	3.10*
University of Michigan math majors [35, p. 71]	1987 -1988	173	26	3.38	3.53*
1st year students at Rutgers in the top 10% of their respective SAT distributions [52, p. 5]	1985	218	47	2.92	3.18*

* Difference tested and significant.

TABLE 2. Grades in mathematics courses
earned by undergraduate males and females.

Course and Institution	N	% Female	Average Grade Males	Average Grade Females	d
College algebra [120, p. 338]	336	52	1.98	2.24[ns]	
Algebra and trigonometry [120, p. 338]	358	39	2.18	2.59*	
Algebra at 6 colleges [7, p. 279]	1,435	48	2.32	2.39*	0.14
Precalculus at 6 colleges [7, p. 280]	3,005	43	2.12	2.38*	0.00
Precalculus[13] [136][†]	3,998	61	2.18	2.32	0.13
Calculus at 8 colleges [7, p. 281]	4,216	35	2.36	2.39	0.14
Calculus [136][†]	21,101	42	2.54	2.68	0.14
Calculus at several universities [46, p. 126]	561	29	2.30	2.77	
Honors mathematics, University of Michigan [35, p. 71]	173	26	3.22	3.38[ns]	
Courses beyond calculus at 51 institutions [171, p. 325]	3,548	33	2.90	3.00	

* Difference tested and significant.   [ns] Difference tested and not significant.

† Details supplied by the Educational Testing Service.

women earn equal or higher grades than men in science and mathematics, as
shown in Table 3 [1, 7, 28, 64, 131]. Examination of the grades earned by males
and females in introductory calculus reveals that females earned about half again
as many A's as males and males earned about a half again as many C's as females
(see Figure 3) [131].

Rates of participation and selection effects.

Do more talented females select mathematics, compared to males? Since females earn higher grades than males in all college majors, the selection would
most likely occur when students were admitted to colleges and universities. As
Leonard and Jiang [63] suggest, some universities may make selection decisions
that underpredict grades for females by relying on mathematics scores on college
entrance examinations. MIT has studied this phenomenon and adjusted admissions criteria to reduce underprediction for females. We return to this topic in

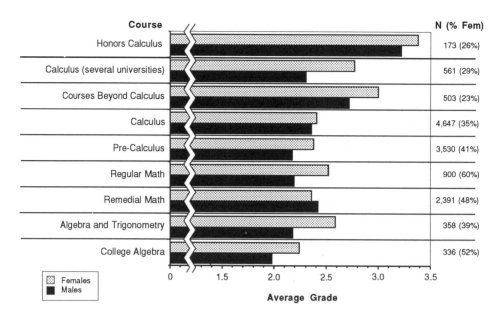

FIGURE 2.   Grades in college mathematics courses.

TABLE 3. Course GPA by group.

	N	% Female		Males	Females
University of Michigan mathematics majors[35, p. 71]	173	26	Math GPA	3.51	3.55[ns]
First year students in 41 colleges and universities [17, p. 21]	165	68	Math and science GPA	2.83	2.98
First year students at Rutgers in the top 10% of their respective SAT distributions, 1985 [52, p. 5]	218	47	Math and science GPA	2.69	2.85[ns]

[ns] Difference tested and not found significant.

the section on test scores.

Institutions vary in the patterns of persistence demonstrated by students. In general, an institution's attention to undergraduate education is associated with higher rates of persistence.

Institutional persistence differences.

As Table 4 shows, the pattern of persistence in mathematics for males and females varies by institution. Overall, 43% of BAs in mathematics granted by

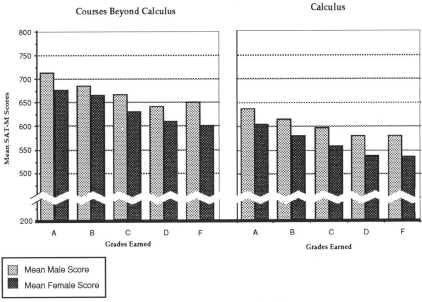

SOURCE: WEINER AND STEINBERG, 1992 [131]

FIGURE 3.   Mean SAT-M scores for males and females earning the
same grade in college mathematics courses.

four-year colleges and universities go to females. Because more men than women
major in engineering and natural science, undergraduate mathematics courses
often have fewer than 40% females. The top ten mathematics departments, rated
on their research reputations, grant degrees to between 9% females (Princeton
University) and 49% females (University of Michigan at Ann Arbor). The average for the top ten schools is 33%. Ph.D.-granting institutions, in general, award
35% of mathematics BAs to women, in contrast to undergraduate colleges, where
women earn 48% of the BAs in mathematics [22].

Looking at research universities, 23 of the 57 top research institutions are
the best producers of women (US Department of Education/National Center for
Education Statistics, [129]). These top producing schools grant BAs to either
50% or more females, or to 40 or more females per year (see Table 5). The top 10
schools' mathematics departments are all top research institutions, but only the
University of Michigan at Ann Arbor is represented among the top producers
of women. Thus, schools with the best reputations in research that have Ph.D.
programs award fewer undergraduate degrees to women than colleges and less
research-oriented institutions.

Among undergraduate institutions, some women' s colleges and liberal arts
colleges have large percentages of mathematics majors compared to the average
of 2% at the top ten research institutions (see Table 6). The total number of
degrees awarded to women by these departments exceeds that of the top ten

TABLE 4. Proportion of females earning BA's in mathematics at all 4 year institutions and at top 10 mathematical research institutions.

A. All 4 year Institutions

	BA's in Mathematics		
Type of Department	N Men	N Women	% Women
Ph. D. granting departments	3,696	1,970	35%
M. A. granting departments	1,933	1,672	46%
Colleges	2,893	2,663	48%
Total	8,522	6,305	43%

SOURCE: CONFERENCE BOARD OF THE MATHEMATICAL SCIENCES, 1990 [22].

B. Top 10 Mathematical Research Institutions
Number of mathematics and statistics baccalaureates awarded in 1991 top ten math departments (research).

	BA's in Math & Statistics			BA's in all fields		
	N Men	N Women	% Women	% BA's Math	Total BA's	% Women
Cal. Tech.	7	5	42%	7%	183	15%
Univ. of Chicago	40	14	26%	6%	857	41%
M. I. T.	40	24	38%	6%	1,107	34%
Princeton	20	2	9%	2%	1,110	40%
Yale	17	9	35%	2%	1,323	44%
Columbia	29	5	15%	2%	1,377	40%
Stanford	13	8	38%	1%	1,470	41%
Harvard	38	10	21%	3%	1,733	42%
Univ. of Michigan	51	46	47%	2%	5,477	49%
U. of Cal.–Berkeley	64	35	35%	2%	5,681	47%
Total	319	158	33%	2%	20,318	44%

SOURCE: US DEPARTMENT OF EDUCATION/NATIONAL CENTER FOR EDUCATION STATISTICS, 1994 [129]

schools. Schools that emphasize the undergraduate curriculum have a higher percentage of female mathematics BAs than other institutions [129].

Participation in precollege mathematics follows a similar pattern. For example, the National Center for Educational Statistics [86] data reveal that between 1978 and 1990 there has been a slight increase in the number of females par-

TABLE 5. All Research I institutions†which awarded a high
percentage (greater than 50) or number (greater than 40) of
baccalaureates in mathematics and statistics to women in 1991.

Location/School	BA's in Math & Statistics				BA's in all fields	
	N Men	N Women	% Women	% BA's Math	Total BA's	% Women
Northeast						
Yeshiva Univ.	4	6	60%	3%	393	44%
SUNY at Stony Brook	64	43	40%	5%	2,164	52%
Boston Univ.	14	25	64%	1%	3,667	55%
Rutgers, New Brunswick	46	46	50%	2%	5,109	53%
Penn. State Univ.	88	82	48%	2%	8,293	46%
Midwest						
Indiana Univ.	29	31	52%	1%	5,123	55%
Univ. of Michigan*	51	46	47%	2%	5,477	49%
Univ. of Minnesota	86	47	35%	2%	5,561	51%
Univ. of Wisconsin	63	43	41%	2%	5,869	53%
Univ. of Illinois	71	43	38%	2%	6,068	46%
South						
Vanderbilt Univ.	29	38	57%	6%	1,169	51%
Howard Univ.	3	5	63%	1%	1,384	62%
Duke Univ.	9	12	57%	1%	1,566	43%
Univ. of Miami	4	5	56%	0.5%	1,820	46%
Univ. of Virginia	19	27	59%	2%	2,815	50%
Louisiana State Univ.	7	13	65%	1%	3,057	54%
No. Carolina St. Univ.	18	27	60%	1%	3,406	40%
Univ. of No. Carolina	29	31	52%	2%	3,538	59%
Virginia Polytechnic	32	35	52%	2%	3,781	43%
Univ. of Florida	30	34	53%	1%	5,498	50%
Hawaii						
Univ. of Hawaii	17	17	50%	1%	2,362	57%
West						
Univ. of Cal.–Los Angeles	97	88	48%	4%	5,105	55%
Univ. of Washington	58	56	49%	2%	5,471	53%

SOURCE: US DEPARTMENT OF EDUCATION/NATIONAL CENTER
FOR EDUCATION STATISTICS, 1994 [129]

---

* Top 10 mathematical research institution

†Research Universities I: These institutions offer a full range of baccalaureate programs, are committed to graduate education through the doctoral degree, and give high priority to research. They receive annually at least $33.5 million in federal support and award at least 50 Ph.D. degrees each year; there were 57 Research Universities I in 1991 [35].

ticipating in advanced mathematics courses, but about 40% of the students are female. Of those taking the AP Calculus exam 43% are females (Blank and Grubel, [4]). In addition, the National Center for Educational Statistics [87] reports that the percent of high school seniors who take eight semesters or more of high school mathematics is stable at about 40% of both males and females. Looking at mathematics courses taken by college-bound students in high school, researchers report that during the four years, females take 3.7 courses on the average, and males take 3.8 courses on the average [134].

What grades do those who switch out of mathematics earn?

Many of the best students in mathematics switch into other fields and more of the best switchers are females. About two-thirds of college students intending to study mathematics or statistics switch to another field. Seymour and Hewitt [108] report on a national sample of over 800,000 students entering four-year colleges and universities in 1987. They found that 72% of females and 60% of males who declare math or statistics majors switch to other fields. Furthermore, of those who switch, half the males and two-thirds of the females switch out of math, science, and technology.

In Seymour and Hewitt's study [108] of 335 students at seven institutions, the mean exit GPA for science and mathematics switchers was 3.3 (range 2.0-3.85) and the current GPA for seniors was 3.2 (range 3.0-3.95). Similarly, Humphreys and Freeland [47, p. 5] studied all first-time freshmen entering the College of Engineering at the University of California, Berkeley in the fall semesters 1985, 1986, and 1987, and found that, "students who persisted and students who switched earned comparable grade point averages (3.10 as compared with 3.07)." The difference was not found to be statistically significant.

A study at the University of Colorado at Boulder offers similar results. For first year students who entered Science, Mathematics, or Engineering majors between 1980 and 1988, the average predicted GPA for those who persisted was 2.93, only slightly higher than the 2.86 for switchers. More women than men switched and the predicted GPA for females was higher for both switchers (female 2.88, males 2.84) and persisters (females 2.95, males 2.92) [108, p. 61].

Furthermore, successful mathematics majors regularly choose other fields for graduate study and females switch more often than males. For example, in a study of career goals of 204 male and female undergraduates at nine Canadian Universities, Nevitte, Gibbins and Codding [90, p. 31] found that the top female students were more likely to switch than average performers.

The best female science students were about five times more likely to seek careers outside science while average performers are about three times more likely than top performers to plan on pursuing further postgraduate education. These data show that the most able female science students are significantly less likely than males to follow up their early success in science with careers within the science community [90, p. 45]. Thus, females earn equal or higher mathematics grades than males suggesting that, if the best students persist,

TABLE 6. Proportion of women earning BA's in mathematics or statistics at selected Women's Colleges and Liberal Arts Colleges.

Women's Colleges (1991)	N Men	N Women	% BA's Math	Total BA's all fields
Mills College	–	3	1%	219
Bryn Mawr College	–	15	5%	295
Spelman College	–	23	6%	383
Mount Holyoke College	–	12	2%	526
Barnard College	–	3	1%	528
Wellesley College	–	10	2%	558
Vassar College	3	4	1%	628
Radcliffe College	–	14	2%	643
Smith College	–	19	3%	741
Total	3	103	2%	4,521

Liberal Arts Colleges (1991)	N Men	N Women	% Women	% BA's Math	Total BA's all fields	% Women
Bowdoin College*	6	6	50%	3%	376	47%
Bates College*	3	6	67%	2%	402	54%
Hamilton College	15	10	40%	6%	425	49%
Middlebury College*	6	8	57%	3%	494	49%
Denison College	3	6	67%	2%	528	52%
Union College*	7	12	63%	4%	538	45%
Saint Olaf College	30	28	48%	8%	708	53%
Total	70	76	52%	4%	3,471	50%

SOURCE: US DEPARTMENT OF EDUCATION/NATIONAL CENTER FOR EDUCATION STATISTICS, 1994 [129]

---

*Dropped SAT as admissions requirement

more of these students should be female. In contrast, large numbers of males and females switch, more of the switchers are female, and those with high grades switch as often as those with low grades. After college, the best female students often opt for fields other than mathematics. To contextualize this situation, we examine the views of students who take undergraduate mathematics courses.

Perceptions of undergraduate mathematics experiences

A synthesis of comments from over 300 persisters and switchers found that between 20% and 35% of both persisters and switchers report that (a) the curriculum is fast-paced, (b) they prefer the teaching in non-mathematics and non-science courses, and (c) they have encountered conceptual difficulties (see Figure 4).

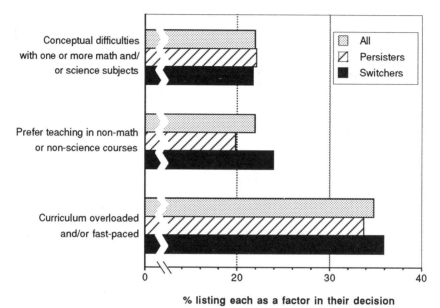

FIGURE 4.   What factors concern both switchers and persisters in math and science?

Three factors distinguished switchers from persisters. First, over 80% of the switchers and 60% of the persisters mention poor teaching in mathematics and science as a concern. Second, more switchers than persisters mentioned inadequate advising and better education in majors outside of math and science. Third, the competitive culture influenced switchers more than persisters (see Figure 5) [108].

The many quotes that follow express students' interpretations of their experiences. The quotes help paint a picture of how students perceive their experiences. We report on mathematics students' learning practices, classroom practices, and teaching practices. In addition, we examine how switchers and persisters perceive academic success, learning environments, career opportunities, and peer and family support.

Comments from male and female mathematics students in a broad range of interview studies help explain both the grades earned by males and females and the reasons many students switch out of mathematics. In this section we report typical comments found in every study of student course perceptions.

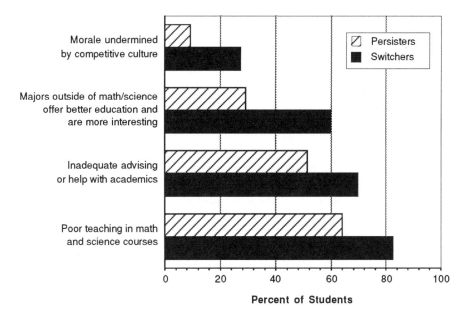

SOURCE: SEYMOUR AND HEWITT, 1994 [108, P. 70]

FIGURE 5.   What reasons contribute to decisions to switch out of mathematics and science majors?

Methodology.

We report on studies of switchers and persisters that use interviews, questionnaires, and, in some cases, written reflections. In all but two of the studies we examined (see Appendix), the students were considered especially well qualified (they had high SAT-M scores or high school GPAs, or attended elite colleges) and were interested in or majored in mathematics or science. Structure, frequency, and duration of the surveys and interviews varied from study to study but all addressed college course experience. Responses were audiotaped and transcribed or recorded in field notes. Investigators coded responses and looked for common patterns. The Seymour and Hewitt study is particularly noteworthy: it is the largest of the qualitative studies we report on (the data set is over 600 interview hours) and by far the most extensive (the 544-page final report addresses a broad range of topics). Its findings are consistent with those of earlier studies of science, and also with studies in fields such as computer science (e.g., Spertus [113]).

These studies describe the full range of student views. Seymour and Hewitt [108] speak of telling students' stories "as faithfully as they know how" and Holland and Eisenhart [45] try to minimize the risk of imposing analyses that their informants might reject by "letting their ideas and feelings come through, in their own words." We too, attempt to convey the views of the many students in these studies by letting them speak in their own words. In these studies, students speak candidly and often their comments sound harsh. At times it is hard to listen to these comments yet they deserve attention as they are, indeed, widely held. Reviewing the range of studies from many different institutions presents a compelling picture. We have selected only comments that are representative of findings across studies.

Learning practices.

The most straightforward explanation for higher grades earned by females is that females learn more in mathematics courses than males, thereby performing better on assignments and tests. For example, a study of males' and females' grades on quizzes and tests in several calculus courses revealed that females consistently earned higher scores [118].

Cognitive analysis of mathematics learning indicates group differences in acquisition of problem-solving methods advocated by the teacher and the textbook. Gallagher [37] found that females more than males learn the traditional procedures for solving algebra problems in pre-calculus courses. Gallagher argued that learning the methods advocated in class was obviously appropriate, but generally meant that the best learners relied on more time-consuming procedures for solving algebra problems. These time-consuming algorithms taught by algebra teachers resulted in correct solutions to algebra problems but could be replaced by "short cuts" that were not taught (see the example in Linn and Hyde [76]). Gallagher conjectured that using these traditional procedures but not the short-cuts might account for some of the gender differences in college entrance examinations. A student comment supports this view,

> "With hindsight, I can understand that my lack of confidence and eventual fear of things mathematical were due in great part to a lack of experience and never having had an opportunity to develop shortcuts for calculations. I tended to try to do mentally what I'd been taught to do on paper and with pencil—borrow from the ten, carry the one, etc.—never rounding off or estimating generally." [139, p. 132]

A variety of studies suggest that on average females follow more effective study procedures than males. In several investigations of self-reported study techniques, females prefer practices likely to yield comprehensive and robust understanding of mathematics (e.g., Linn [69]). In a study of college computer science courses, females report spending more time reflecting on the similarities among problems they study, more time organizing and linking their ideas, and more time planning and reviewing material for problem solutions [80]. Overall, females' self-reported study practices reflect a goal of gaining an integrative view of the field they study.

To explain these findings, we draw on reflections of female engineering students from Princeton. One student reports,

"I did notice, however, that the women tended to work together on homework. This seemed to be because we worked in the same style. The two
or three women I worked with the most usually started the problem sets
early, then compared them with each other. This left time for going to see
the professor if we were still stuck after consulting each other. It would be
interesting to know if the professors notice whether (as a percentage) more
women came to them for help. I know that one or all of us were in the
professor's office every week. Lots of the men I know started the problem
sets the night before they were due and just did their best and handed
them in. We (my women friends and I) almost never turned a problem set
in that wasn't perfect. I don't know whether these are good generalizations
or if I simply attracted people like that around me because I am that way."
[91, p. 69]

Classroom practices.

In mathematics classes, instructors ask males more questions than females,
expect males to answer more difficult questions than females, and give males
more time to respond than females. In addition, males ask more questions than
females and call out or interrupt more and participate more in discussions [57,
58, 61, 100, 101, 102, 133]. As a result, males participate more in the discourse of
mathematics than females. More importantly, however, females begin to define
themselves as observers and to conclude that the standards of discourse put them
at a disadvantage. Females who adopt the strategy of aggressively entering the
fray may be labeled "femi-nazis," patronized with instructions to review the
material more carefully before speaking, or stereotyped as unworthy of a reply
[30, 41, 53, 122]. These discourse practices can reinforce insecurities of female
students. For example, one student reports:

"I am the type of person who feels insecure no matter how well I do in my
classes. I am always thinking that I don't know as much about computers
as everyone else in the department, no matter what their GPA. I have also
convinced myself that my independent work project is not as difficult or
significant as other people's. During sophomore year, I was a basket case.
I thought I knew nothing and that all those nerdy guys in my classes were
geniuses or something. After a while, though, you figure out that they're
just as stupid (or smart) as everyone else." [91, p. 37]

Females in mathematics classes become more and more silent. Although they
comprise 40% of the students in mathematics classes, on average, it is common
for both males and females to estimate that fewer than 20% of their classmates
are females [55]. Females learn the material and earn higher grades than males,
but males define the rules of discourse and females observe.

Teaching practices.

College and university students often report disappointment in introductory math courses, especially compared to high school calculus experiences. In high school, their calculus teachers viewed themselves as fortunate to instruct the best students and to teach the most advanced material. College calculus instructors view themselves as teaching a broad range of often unprepared students and as covering routine material. Recent efforts to reform calculus instruction attest to the past neglect of those courses and offer promise for the future [14, 20, 29, 128].

At many institutions, students report a few examples of college mathematics lecturers who read directly from the textbook. Seymour and Hewitt [108] found students on all seven campuses they studied who reported this practice. These comprised a cross-section of institutions: a small private liberal arts college, a private city-based university, a large private university on the West Coast with a highly selective admissions policy, a multi-role public urban university in the Northeast, a large urban public university in the Midwest, a state university in the Southwest, and a large state university considered the "flagship" institution for its southwestern state. They report interviews with 335 science and mathematics majors, all of whom had SAT-Ms of at least 650. Here we give a flavor of the comments of students from these various institutions. One male engineering student reported,

"I had one professor who would literally pick up the book and read it to the class. I mean, he would just read. He was actually not bad for the math department, but we had 60% of the class drop the course. I counted one day, and out of maybe one hundred eighty students, seventeen showed up for the lecture. I was happy with the course content, and the facilities were wonderful, but the teaching was just a vast disappointment to me." [108, pp. 215–216]

A female student describes another frustrating situation,

"They just continuously write. And they're standing in front of what they write, but just don't care. And they'll look over their shoulder now and then, and say, 'Okay, you all are still there.' And then they just keep going. And the number of people that don't go to classes is amazing here, truly." [108, pp. 215–216]

Some students who persist in mathematics may do so for altruistic reasons hoping to change the way mathematics is taught. Thus one female who persisted reports,

"I am, hopefully, going to teach in a completely different manner from my own math teachers, just because they were all so boring. I want to make my classes fun, so that kids will enjoy math." [108, p. 86]

Other female persisters say,

"There's a lot of times I tried to find professors in their designated office hours, and they were just not to be found. And you'd try to catch them after class, and make an appointment with them. It got really aggravating." [108, p. 183]

"Part of the problem with the math department, I think, is their attitude. I think they realize they're bad, but they don't really care. It's not their problem that their students are failing their courses. It's the students' problem." [108, p. 209]

Another study of undergraduate mathematics and science programs on five campuses, each with a strong commitment to teaching, yields similar comments. At all five institutions professors rather than graduate teaching assistants teach basic courses and tenure and promotion depend on effective teaching. Yet, Astin and Astin [3, p. 102] report:

"In spite of this commitment to teaching, very few innovative teaching practices were employed. The vast majority of the courses at all the institutions we observed were taught in a traditional lecture style. The professors typically stood at the front of the room, often behind a table or lectern, and used a chalk board or overhead projector to illustrate a concept or write out a formula. Although all of the professors we observed had Ph.D.'s and were considered experts in their fields, few of them seemed able to present their knowledge in an interesting or provocative way. Many professors mumbled, avoided eye contact by looking at the floor, and asked rhetorical questions that they quickly answered themselves. The 'energy' was very low in such classes. Many students arrived late and many left early. In essence, students were not engaged in the learning process. Some students expressed disappointment with the 'boring' lectures of certain faculty who taught the same courses year after year. A physicist at Santa Clara, however, defended the traditional lecture method: 'You get it alone by thinking hard . . . That's what I had to do . . . That's what they're going to do!' Nevertheless, most students seem to accept traditional teaching methods. We believe that this acquiescence occurred in part because of their lack of exposure to any other teaching style."

In another study, a computer artist comments on a required calculus course:

"I took calculus because it was required for computer science. I was frustrated by the professor—disappointed by his method of teaching. He mumbled. It seems to be the way that overcrowded departments work. They hire a professor for a different purpose than teaching. They are here to do their masters or are writing a book." [79, pp. 167–8]

Academic success and switching.

Virtually all studies show that those who switch have grades and preparation comparable to those who remain in mathematics [107]. For example, one student comments,

> "My first year I did really well—like 3.8. And it was a surprise to people, like my mother, that I wanted to change my major—because I was doing so well. And I told her, 'It's not that it's hard. I just don't want to do it.' After that, my grades dropped off a bit, because I wasn't sure what I wanted to do—and that was purely through lack of interest. Last semester, after I had officially changed my major, they went back up to a 4.0." [108, p. 148]

Learning environments.

Switchers out of mathematics most often complain that the learning environment drove them away [108]. They describe discouraging experiences in math classes, non-existent faculty advising, and stereotyped peer pressures. In many cases, efforts to weed out students discourage the wrong students by creating an unrealistic image of mathematics. For example, a student in yet another study commented:

> "In college I started Calculus I. I could understand all of the concepts as it was mostly a review of high school [calculus]. However, this is a known 'weeder' course. I felt, even though I knew the material, that there was no way I could compete with the pre-med students and engineers. I became very stressed. After the first exam I dropped the course, never again to take another math class. . . .
>
> I think that a real help would have been in college, if there had been less emphasis on 'weeding' people out of math and more on keeping people (especially women) in. My T.A. [teaching assistant] did nothing to keep me from dropping (not even a word). I think if a place like [my university] had all women's sections with women T.A.s, they could make women students feel more comfortable and perhaps more would pursue math and science courses and careers." [139, pp. 149, 151]

Other students commented:

> "I always liked math, and I was still good at it, so I was thinking maybe I'd integrate it with business. Then, after looking at it some more, I had to confess that I'd lost much of my interest in it, and I changed to geography." [108, p. 254]

> "Their attitude is that they don't expect you to make it through. It's very discouraging. You know that doesn't encourage you to do your best. I felt

they were telling me, 'No you can't do it. You're not going to make it'."
[108, p. 182]

"I expressed an interest in math when I came here, and I had him as a
professor last year, too, but I haven't got to see him as my advisor once
this year. And, it's not like he has hundreds of students to advise–the most
they have is about twenty-five. You'd really think it wasn't too much to
ask to come in once and a while and talk with you a little bit." [108, p.
197]

"I never knew exactly whether or not I was playing by the correct rules,
because it seemed like I talked to one person, and they'd say something;
and I'd talk to someone else and they'd say something else.... And they
definitely need to get their T.A.s advised about how the system works. I
mean, even the dean didn't know what to do." [108, p. 191]

"I had a horrible math professor freshman year. He didn't speak any English and he was failing everyone. He turned me off. I went to an advisor in
the math department and he was of no assistance either." [80, p. 167]

Another study reports similar comments.

"Students are influenced in their course choices and their concentration
decisions by many people: peers, parents, teachers and advisors. But this
advice is often remarkably unhelpful, especially when a student has no way
of evaluating it or of putting it to use in an internal process of decision-
making. [One student comments:] 'I took the math placement and I was
on the borderline of Math 1a and Math AR, and they said "Take Math
AR," so I did that. [Interviewer: "Who said that?"] The gods from above
who sent back the computer readout from the placement test'." [124, p.
75]

For some students, poor teaching leads to switching but for others it does not.
Students describe difficult and often frustrating decisions (see Figure 6).

Career opportunities.

Another aspect of persistence concerns career opportunities. Females often
mention limited career options for mathematics majors as a problem. For example, one Native American female who switched out of mathematics commented,

"I talk to some of my friends that are math majors right now, and they're
saying, 'What am I gonna do.' I mean they look to grad school because
they don't know what else to do. . . . People say you can do so much with
a math major—the problem is finding it." [108, pp. 260–261]

I always loved math, but, all of a sudden, I'm coming into sophomore year, and I started thinking about what I was going to do with my life . . . And people start asking what I was going to do with a math major, and where I was going to work, and I had no idea.

—White female, switcher to social science [108, p. 36]

I always loved math, but, coming into the sophomore year, you know, I started thinking more about what I was going to do with my life. . . . And people kept telling us there were all sorts of things you could do with math, but they never told us what these were.

—Female white mathematics switcher, SAT-M over 650 [108, p. 260]

I didn't know that Princeton had engineering when I first applied. I was thinking of math but randomly found myself at an electrical engineering orientation meeting. . . . Why I choose EE over, say, mechanical was because I liked math more than physics and the relation of EE seemed stronger to math.

—1993 Princeton engineering graduate [91, p. 52]

Originally, I was going to go into mathematics; that's why I chose Princeton over Urbana-Champaign. It's sort of random how I got into electrical engineering: I met a math major who completely intimidated me, and I thought if everyone in the class was like him, I should try something else. I had heard that electrical engineering was very mathematical, so I thought I'd try it and see what it was like.

—1992 Princeton engineering graduate [91, p. 38]

I see a lot of people here who have gone to college" while still in high school, taken courses like this one already. They are math majors because they are two or three years ahead of me. How good a math major am I going to be if I am [already] two years behind?

—Harvard-Radcliffe student [124, p. 77]

The grapevine has it that people don't major in math or physics here unless they were child prodigies to begin with. This isn't to say that if I had some overriding desire to do math or physics I wouldn't. But it is something to be overcome.

—Harvard-Radcliffe student [124, p. 78]

Even though I had done really well in the math courses I took here . . . one of my section leaders told me that only people who start out in Math 55 keep taking math.

—Harvard-Radcliffe student [124, p. 78]

It is really quite a shock when people are so competent. I went into Math 1A first semester, not prepared . . . and I had to drop back because of the kids who are majoring in science.

—Harvard-Radcliffe student [124, p. 78]

FIGURE 6.   Comments of students who switched out of mathematics.

Other female switchers saw the options as basically dull,

"You start broad, and now you're just narrowing down your life to this straight path going somewhere. But no one ever gives you a clue about what kinds of things you're supposed to be doing out there." [108, p. 261]

"I kinda looked at a couple of jobs, but I realized, no matter what I did with math, it's pretty much going to be a nine-to-five job in an office. Right away, this was my biggest turn-off. It's not so much math; I still love math." [108, p. 86]

Jackson [49, p. 448] reports similar conclusions raised by minority students:

"For many reasons, including those I just mentioned, minority students are attending medical, business and law schools instead of graduate schools in mathematics. The unfortunate thing is that some of these students actually prefer mathematics. However, to many gifted minority students with several career choices, mathematics is not a good choice. Many view mathematics as having too many roadblocks, an area where they are unwelcome, where there will be limitations. Let me give you an example. I know of a student who came to Hampton University with a very good SAT score and graduated at the top of her class as math major. She did well in every program she was placed in, including a summer program at Ohio State University. At the conclusion of her studies at Hampton, she told me quite frankly that she preferred mathematics, but from what she had seen of the mathematics profession, she would be better off in another profession. She is now a medical student at Duke University, exploring the possibilities of somehow satisfying her appetite for mathematics."

Peer and family support.

In contrast, a broad range of evidence concerning females who do choose to persist in mathematics reveals the tremendous support that they get from their families and peers. For example, one student from Princeton comments on how her family helped her persist in a traditionally male field:

"The one thing that I really like about being Chinese-American is the close-knit family ties. I think that played a major role in my being able to get into here. My family was very supportive every step of the way. Even after I graduate they will be behind me 110.2223 %. I tend to be overprotected, but I know they mean well because they love me. They're ready to let go. A lot of times I'll talk to them about a problem and they'll say, OK, you're old enough to deal with it by yourself." [91, p. 40]

Another female who persisted in science reports:

"My sister's a lawyer, and another sister's a C.P.A., so we're a very career-oriented family. Even going back to my grandfather who worked on the railroads, and you might think it would be the opposite. He, too, would say no to women just staying home and caring for children. 'You need to graduate from college,' he would tell us from being little. And so there was no question in my mind from day one that that's where I was going." [108, p. 378]

My godmother was an engineer from MIT when there were few female graduates from that institution and she encouraged me [to go into engineering]. I had an early interest in math, but decided to study engineering because I thought it was not as "academic" as pure math.

[91, p. 28]

I suppose because I left engineering, I think more of having been an "engineer" rather than being a "woman engineer." However, reflecting back on that time, I think I definitely experienced being a woman engineer. I decided to study engineering because my father was an engineer. In fact, I chose chemical engineering because my father was a chemical engineer; I was always surprised at how many other women engineers had a parent who was an engineer. For women, it always seemed that they had a reason for studying engineering—"I'm good at math and science." For the male students, studying engineering was never even questioned. If they shrugged and said they were studying engineering because they did not know what else to study, this reason was readily accepted. I don't think a woman could have gotten away with that, and that's significant.

[91, p. 42]

You miss someone to have camaraderie with. The guys would all be laughing and talking, and I'd sit there in the corner with my math book on my own. It wasn't until the 400 level classes that there were other girls in the class. And it was much nicer after that. Two or three of us would sit together, and have someone to talk to. Actually, it was out of class too—I mean we'd laugh and joke and entertain each other during class, and help each other with small things we'd missed. And if we didn't understand something the professor said, we'd huddle and check it out with each other.

—Female white science non-switcher [108, p. 398]

I talked with my advisor a lot, and he would encourage me when I faltered. He'd say, "Look at your good grades. You can do this!" He was backing me up right from my own performance. I saw that he had the same kind of relationship with some of the guys, too-just that kind of mutual respect and support.

—Female white mathematics non-switcher [108, p. 406]

FIGURE 7.   Comments of female mathematics students about peer and family.

Implications of student perceptions.

Overall, reading these comments from students, it appears that females earn higher grades than males in high school and college mathematics courses because they have better study habits and learn more mathematics. In these courses, however, students learn more than mathematics. They learn to participate or to be silent. They observe the discourse rules of the field. And, they learn that introductory courses are designed to filter out rather than to interest, teach, or nurture students.

Although students who leave mathematics are similar to those who persist in terms of grades, high school preparation, and college entrance examination scores, many faculty believe that the best students stay and, "people whose mathematical past has caught up with them and should have been flunked out earlier" [84, p. 7] are the ones who switch. In fact poor teaching and advising may select for the most determined rather than the most talented students. Research universities and institutions with Ph.D. programs appear to discourage more women than liberal arts colleges. Neglect of female students may account for the larger number of female switchers. Courses intended to filter out students may deter the best students and retain the most stubborn.

Patterns of persistence and reform of the curriculum

In response to complaints from students, faculty dissatisfaction with traditional calculus courses, and better understanding of mathematics learning, exciting calculus reform efforts are now widespread. These reforms also reflect agreement that calculus courses, as currently taught, do not meet the needs of many students. In particular, many believe that the exercises in standard textbooks are far too narrow. Traditional calculus courses emphasize routine application of algorithms, fleeting coverage of numerous topics, and piecemeal approaches to complex concepts. As the comments from students suggest, these courses make it difficult to connect and link ideas, recognize fundamental concepts, and see the relevance of mathematics to work in other disciplines. Helping students develop their own problem-solving processes requires analyzing the problem-solving process and illustrating techniques for identifying patterns and detecting errors.

Course projects can respond to concerns raised by students in the last section [16, 68, 104]. Projects have proven successful in several undergraduate programs and are a feature of "lean and lively" calculus [19, 29, 115, 126]. This approach typically also includes fewer topics discussed in greater depth. It may feature technological tools such as *Mathematica*, *Maple*, or *MathCAD*.

To help students learn from projects many reforms of undergraduate courses and especially calculus emphasize projects combined with support for students as they engage in sustained, complex mathematical thinking [20, 104]. The process of supporting students as they learn to think mathematically has been called "scaffolding" [9, 21]. Reformers recommend that scaffolding include emphasis in discussions on alternative solutions, wrong paths, methods for detecting errors, and recyclable patterns. This approach goes well beyond typical instruction where instructors tell students the correct algorithms and then assign problems that students view as routine and disconnected from other parts of mathematics.

Creating good project assignments challenges faculty. Students might be asked, for example, to model the rate at which an audience leaves a crowded theater (see other examples in Figure 8). Often these projects require students to work with their peers.

Implementing calculus reforms including course projects necessitates communication to students about the changed goals and activities, and communication to the broader community concerning the expected skills and knowledge of stu-

If a_1, a_2, a_3, . . . are the positive integers whose decimal representations do not contain the digit 7, show that

$$\sum_{n=1}^{\infty} \frac{1}{a_n}$$

converges and its sum is less than 90.

Here are some suggestions: Break up the sequence of positive integers into carefully chosen blocks (not necessarily all of the same size), count the number of terms in each block, and see what fraction of them is left after removing those with an offending 7. Then bound the size of the sum of the reciprocals in each block, and try to compare this result to a convergent series.

[20, p. 195]

An ant at the bottom of an empty sugar bowl eats the last few remaining grains. It is now too bloated to climb at a vertical angle as ants usually can; the steepest it can climb is at an angle to the horizontal with a tangent equal to 1. The sugar bowl is shaped like the paraboloid, $z = x^2 + y^2$, $(0 \leq z \leq 4)$ where the coordinates are in centimeters.

(a) Find the path the ant takes to get to the top of the sugar bowl, assuming it climbs as steeply as possible. Use polar coordinates (r, θ) in the xy-plane, and think of the ant's path as parameterized by r; then find a relation between the differentials $d\theta$ and dr, and integrate this relation to get $\theta(r)$.

(b) What is the length of the ant's path from the bottom to the rim? To answer this, first discover a formula for arc length involving dz, dr, and $d\theta$ in three dimensions.

(c) Draw a graph of the sugar bowl and the path the ant takes to get out.
    HINT: You may want to start with the projection of the path in the $r\theta$-plane.

[20, p. 208]

The alternating harmonic series $\sum_{n=1}^{\infty} \frac{(-1)^{n+1}}{n}$ converges to $\ln 2$.

Now suppose s is any given number. You are going to prove the amazing result that the alternating harmonic series can be rearranged (that is, its terms can be written in a different order) so that the resulting series converges and has s as its sum.

You will need to understand the precise definitions of the limit of a sequence and the sum of a series.

HINT: Start by trying to reorder the terms to alternately overshoot and undershoot s with various new partial sums.

[20, p. 167]

FIGURE 8.   Examples of calculus projects.

dents taking the new courses. Naturally students and faculty often resist these changes. Experience introducing projects in several courses has convinced us that a framework is needed to guide this change. Both students and faculty need scaffolding to make this new approach effective. A possible framework is described briefly in the next section.

Reform is not easy. (In the next section we give a brief discussion of research whose implications might make it easier.) Students' initial reactions to reform are often negative [18, 54, 105]. At the University of Michigan students complained that the "New Wave" calculus was unsettling and painful [8]. They said that courses took too much time, were confusing and ambiguous, ("How can we tell if we have the correct answers?"), and offered poor preparation, ("I know calculus and this isn't it!," "I won't be prepared for the next course," "Why are we having to do all this writing? Writing has nothing to do with mathematics!"). Group work was frustrating, ("It isn't fair for my grade to depend on the work of others").

Consider student reactions to a reform project at Duke. One student wrote, "I wish I had to memorize more. I'm sick of real life models" [24, p. 1061]. Duke first-year student Mary Harris, who likes math and is considering becoming a math major, sums up the reform this way: "It's a big exercise in confusion." When asked a question, instructors typically responded with another question. Students have trouble finding ways to learn under these new conditions [24, p. 1061]. In fact, the students who protested the most did the best under the old system. "These kids have been really successful at high school, under the old rules," says Jack Bookman, a Duke math instructor who is evaluating the Duke program for NSF.

Many schools abandon reform in the face of student resistance [15, 18]. Positive results have led others to persist. For example, group work appeals to some students, "Group work gives us more understanding—helps our learning," "Homework groups make us do and understand the work!" [8, p. 1]. Others report that over time, students see benefits. "I believe I learned more in this class than I ever learned before," wrote one student [24].

Even if the students eventually recognize the advantages of reformed courses, faculty often resist. "Without compelling reason to change most teachers will stick with traditional methods—after all, those methods produced such fine upstanding mathematicians as themselves" [15]. Of course, some faculty have endorsed the approach. One remarked, "It's amazing how much one's estimate of students goes up after using projects" [20]. The ACRE study [125] shows that how calculus is taught has changed more than what is taught. Changes in instructional practice, more frequent use of technology, and increased focus on building students' conceptual understanding with less attention to symbol manipulation, are finding their way into calculus and other mathematics courses.

In summary, changing calculus courses requires serious attention to helping faculty scaffold students and helping students become independent learners. Nonetheless, change is underway. Recently, 80% of the 62 departments responding to an in-depth survey of calculus teachers reported using graphing calculators, 65% employed modeling and applications more heavily, and 40% reported increased student project activity [125].

The scaffolded knowledge integration framework.

To successfully reform courses, students need to learn new skills including those required for long, complex projects and faculty need new approaches for communicating to students [14, 20]. Students prefer traditional problem sets to projects because they have succeeded on these in the past [66, 67]. Students often fail to collaborate when working in groups because they have no prior experience with this form of learning [2]. We describe tactics for introducing reform from a variety of studies to illustrate promising practices. We focus on projects rather than all aspects of reform and on one framework that has succeeded. Efforts to introduce projects into a middle-school science class, Pascal and LISP courses at the university, and an engineering design course at the university resulted in a framework called scaffolded knowledge integration. In these studies several cycles of trial and refinement were used to improve the course and establish new performance standards [72].

As noted above "scaffolding" is a process of supporting students as they learn to cope with new and complex tasks [9, 21]. Many reforms of undergraduate courses and especially calculus emphasize projects combined with scaffolding to allow students to engage in sustained, complex mathematical thinking [20, 104].

The scaffolded knowledge integration framework starts by defining new goals for instruction that meet student needs and provide an authentic experience of the field. The projects advocated for calculus fit this aspect of the framework. Reformed calculus courses have a new goal of engaging students in authentic mathematics activity. In addition, the scaffolded knowledge integration framework emphasizes the need for balance between what could be called "making thinking visible" and "supporting autonomous learning." To make thinking visible, courses illustrate the steps necessary to complete sustained investigations. If the steps are too constrained, projects look just like problem sets. The goal is to help students learn to solve more complex problems on their own. What sort of assistance makes this possible? First, students benefit from models of the outcome expected in order to understand the assignment that they have been given. Providing students with model calculus projects completed by students in previous semesters or at other institutions can help improve the quality of the projects and the comfort level of the students. Case studies that communicate the process of problem solving, not just the solution to the problem, can also help students identify strategies for solving large, complex problems. Case studies in computer science, engineering, and business help students complete large projects by making some of the steps necessary to complete projects explicit and by illustrating how to organize information in reusable patterns [72, 74].

Ultimately students need to solve problems on their own, or what we call autonomously. They need to become independent learners. If too much help is given, they get no practice in figuring things out themselves. Yet, without help students may endlessly flounder or produce unsatisfactory work. One success in the scaffolded knowledge integration framework was to encourage students to criticize the solutions of others. Those who acted as critics learned both how to solve problems and how to analyze potential solutions, while those who only solved problems were prone to accept incorrect solutions even when they were able to generate correct solutions themselves [26].

To help students work in groups, the scaffolded knowledge integration framework also emphasizes providing social support for learners. All learning takes place in a social context, so the goal is to structure social interactions to support all learners. Initially, groups often have difficulty working jointly on problems [19, 73]. Sometimes one student does all the work. Often students complain about the lack of contributions from their peers. Students participating in groups may reinforce stereotypes concerning who can contribute to a particular field. For example, in engineering, Agogino and Linn [2] report on groups that convince women students that their contributions are less valued than those of male students or silence students who could contribute. Mathematics faculty and former undergraduates report similar results [91, 106]. Group work can help students develop collaborative learning skills when accompanied by effective instruction. Modeling collaborative learning skills and judiciously designing group experiences makes for improved learning.

So far, research suggests some clues about how groups can help each other learn. For example, taking the role of a tutor in a group can help students recognize the limits of their own understanding. Students who tutor others in groups frequently learn more mathematics than students who work on their own [103, 132]. Furthermore, when students use each other's ideas they learn how to make sense of disparate information. When they rely on texts to suggest ways to think about problems, students often take the text as the authority, but they are likely to analyze ideas from their peers. Sometimes students can describe a problem in words that are understandable to other students when the textbook or faculty fail. This reinterpretation of mathematical ideas helps students make sense of the material [73, 111].

Overall, calculus projects have the potential to communicate the design and discovery activities of mathematics, and help students link and connect ideas. By adding projects and modifying instruction to help students complete projects, calculus instructors can enhance understanding of the nature of mathematics. Projects allow students opportunities for deep understanding that do not arise in courses with only short problems. Traditional courses may convince students that mathematics is algorithmic and boring, thus discouraging talented students and especially women [10]. Project-based courses that feature appropriate support for students may attract the most creative individuals. These project experiences allow all students to engage in activities common in undergraduate research. Often only the best students, as well as far more men than women, have the opportunity for undergraduate research [107].

Designing effective courses typically requires several cycles of innovation and revision. Mathematics faculty can speed this process by communicating their own experiences with reform and learning from each others' successes.

Mathematics college entrance examinations

If women get higher grades in mathematics, why are they earning lower scores than men on college entrance examinations such as the SAT-M and the ACT mathematics? How are grades in mathematics courses related to scores on mathematics aptitude tests? How do these tests fit into the plans for calculus reform?

General tests of mathematical ability administered to representative samples of precollege students show no gender differences in performance [12]. For example, Hyde, Fennema, and Lamon [48], performed a meta-analysis of hundreds of studies conducted in the United States and found an overall effect of about one-tenth of one standard deviation. Internationally, in some countries females out-perform males while in others males out-perform females [34, 39, 43]. However, college entrance examinations reveal a different picture.

We analyze college entrance examinations from the standpoint of their validity as predictors of performance in college mathematics courses. Since these tests require students to solve 25 to 35 problems in 30 minutes, there is a premium on efficiency. In contrast, the reform of mathematics courses, to include an emphasis on solving complex ill-posed and personally relevant problems, places a premium on sustained reasoning. The rapid solutions to problems characteristic of college entrance examinations are a minor component of calculus projects. And, ability to refine complex problems and design appropriate solutions goes beyond the skills measured in college entrance examinations. Thus, at least on the face of it, the validity of college entrance examinations for predicting performance in reformed versions of the calculus curriculum is likely to decline.

As discussed in the next section, grades earned in these new courses may help determine which students will succeed in more advanced mathematics courses and change the criteria that govern student decisions to switch out of mathematics. Students taking these reformed courses are likely to be better able to make decisions concerning continuing in mathematics because their experience with mathematics will be more authentic.

Before most students enter these courses, however, they must take college entrance examinations. These examinations convey a different view of what constitutes a successful mathematics performance. In addition, males out-perform females on these tests, reinforcing the view that mathematics is a male domain [70].

Performance on SAT-M and ACT-M.

The voluntary sample of college-bound high school students taking the SAT-M or ACT-M shows a consistent gap between male and female performance of about 0.4 standard deviation units [76]. This gap has narrowed slightly over the years but remains large [36].

In the 1980's score differences could be statistically eliminated by adjusting for high school course experience [12, 137]. Examination of recent data from the Educational Testing Service, indicates that for students with the same course experience, the performance gap on SAT is about 0.4 standard deviation units [40, 69]. Recently, the gap in course experience has narrowed but the score gap remains [131]. There is little relationship between SAT scores and performance in all levels of college mathematics. For males there is a slightly greater relationship than for females. Females need about 50 fewer points on the SAT to perform as well as males in the full range of college courses (see Figure 3).

This pattern holds in studies of high school students as well. For example, Gross [42] studied more than 4,000 high school students in Montgomery County, Maryland. Girls took the same advanced math courses as boys. These included calculus, pre-calculus, and advanced algebra studied in the same classrooms and with the same teachers. Girls earned higher grades but their SAT scores were lower by about 0.4 standard deviation units.

Many studies suggest that the multiple choice, rapid response context of SAT and ACT achievement tests favors males over females. The sustained reasoning and organized communication required for essays seems to favor females [69, 88]. Furthermore, mathematics examinations in the Netherlands, England, Australia, and other countries that require solutions to several long problems seem unbiased with regard to gender [27, 39, 60, 85, 117]. Thus, the context of the multiple-choice achievement test may differentially advantage males.

The speeded nature of the college entrance examinations also probably advantages males, who tend to be more confident about their ability to solve problems than females. Studies suggest that males, compared to females, expect to answer correctly, independent of actual performance [75]. This perspective results in less answer checking and reflection, useful skills for speeded tests. Similar findings about speed on spatial reasoning have been reported for males and females [77].

Are these achievement tests valid measures of mathematics ability and potential? Examples of questions that appear on the SAT items are shown in Figure 9. These questions typically distinguish students scoring in the mid-range (500-600). The SAT requires students to solve 35 problems in one 30-minute section and 25 problems in the other 30-minute section. Scoring involves counting about 10 points for each correct answer, minus a percentage of incorrect answers. As a result, a difference of about three to four correct answers accounts for the gender difference. Thus, students must solve problems quickly, pace themselves appropriately, and maintain concentration for a reasonable period of time. Students report that skills such as eliminating the answers that are known to be wrong, using short-cuts and rules of thumb to solve the problems, and detecting trick questions contribute to high scores [97]. Some students view the test as more a measure of trick-detection than a measure of mathematics skill. Females report this as one more reason to conclude that math is frustrating rather than interesting.

Does this test measure mathematical abilities relevant to college work? The most complex items on these tests tap some of the skills that students need to be successful in mathematics yet few students even attempt those items since they are at the end of each section. The concept of a function is beyond the scope of this test yet central to college work. Rapid responses may contribute more to success on these tests than the problem-solving skills and mathematical knowledge students need for future courses.

Do these tests predict undergraduate success? The narrow range of performance among students accepted in the most selective colleges makes the relationship between performance on these tests and undergraduate success relatively limited. For example, a study at MIT found that aptitude test scores account for only 5 to 7% of the variability in grade point average by the end of the sophomore year [50]. A new version of the SAT, out this year, addresses some of these concerns but preliminary data show no change in the gender gap [11].

Using tests in college and scholarship decisions.

Although aptitude tests account for a small percentage of the variance in college grades, their impact on persistence of men and women in mathematics is far greater. When selective scholarship and admissions decisions are made primarily on the basis of aptitude test scores, then two-thirds of those selected are men. Conversely, if decisions were made primarily on grade point averages more women would be selected. These scores underpredict grades for women and overpredict grades for men. Testing organizations recommend balancing scores with other, more comprehensive indicators yet these tests continue to reinforce stereotypes about mathematics as a male domain.

Scores underpredict the ability of women to succeed. Historically, women have earned higher high school and college grades than expected on the basis of their scores on ability or achievement tests [130]. Originally, Thorndike labeled this "over-achievement" [59, 123]. Research investigated the effect of motivation and study habits on grade performance. More recently, the relationship between scores and grades has been labeled "underprediction" [96, 119]. Attention has shifted to the psychometric characteristics of these tests, and especially to their validity in explaining this underprediction for females.

For example, Leonard and Jiang [64] report a study of the cumulative grade point averages at graduation of the approximately 10,000 students who were admitted to the University of California, Berkeley as first year students between 1986 and 1988. As expected, females earned higher grade point averages than males. Some of this difference reflects the majors that males and females choose. But, when field of study is controlled, the SATs continue to underpredict women's cumulative college GPA by a small but significant amount. Had women with lower SATs been selected instead of the men at the cutoff point, the women would have earned higher GPAs than the men who attended the university. The

372	434	124
62	320	558
496	x	248

10. In the large square above, the sums of the numbers in each row, in each column, and in the two main diagonals are all equal. What is the value of x?

   (A) 181   (B) 184   (C) 186   (D) 188   (E) 190

11. A certain number x is multiplied by 6. The number that is 5 less than x is also multiplied by 6. How much greater is the first product than the second?

   (A) 5   (B) 6   (C) 25   (D) 30
   (E) It cannot be determined from the information given.

11. When the sum of two numbers r and s is subtracted from twice the difference of r minus s, the result is equivalent to which of the following expressions?

   (A) $2(r - s) - r + s$ (B) $2r - s - r + s$ (C) $2r - s - (r + s)$
   (D) $2(r - s) - (r + s)$ (E) $2r - s + (r - s)$

12. The daylight period is defined as the time between sunrise and sunset of the same day. If sunrise was at 6:39 a.m. and sunset was at 4:47 p.m. on a certain day, at what time was the middle of the daylight period?

   (A) 11:30 a.m. (B) 11:43 a.m. (C) 12:00 noon (D) 12:13 p.m. (E) 12:24 p.m.

14. If the product of two consecutive positive integers is 5 more than their sum, then the lesser of the two integers is

   (A) 1   (B) 2   (C) 3   (D) 4   (E) 6

FIGURE 9.   Examples of SAT items.

use of SATs in the selection formula resulted in selecting males who earned lower grades in their undergraduate majors than would have been earned by the females who were rejected.

Recent examination of data like these has resulted in modification of admissions standards at many colleges and universities. For example, MIT has dramatically reduced score requirements for the SAT mathematics section, resulting in the addition of many women to the MIT class and a narrowing of the gap in GPA between males and females [50]. Similarly, Rutgers University has examined scores and grade point averages of males and females and modified policies to increase fairness [99]. An investigation at Princeton University revealed that SAT scores of the Princeton class of 1990 were higher in mathematics for males than for females. In spite of these different SAT scores, women's grades were higher than men's and the most significant underpredictor of grades for women was the SAT math score [99].

A number of institutions have reduced or dropped use of entrance examinations like the SAT and ACT (see Table 6). Bates in Maine, Bowdoin in Maine, Middlebury College in Vermont, and Union College in Schenectady, New York have all dropped use of the SAT altogether [13, 99, 110]. At Penn State the influence of the SAT on college admissions has been reduced in recent years [93]. Analysis of applicants who did not submit SAT scores to Bates College reveal that those omitting the SAT scored 80 points lower than applicants who submitted their scores, but did not differ significantly in first year GPA or academic standing [93].

Organizations, especially those concerned with equity for women have argued against the use of SAT scores to award scholarships based on high school achievement [23]. Legal proceedings arguing against the use of the tests for awarding of scholarships have had some success [114].

A particularly inappropriate use of these college entrance examinations scores is for placement in college mathematics courses. Innovative programs at several colleges use either a repertoire of indicators or rely on students to make sensible decisions themselves. Both empirical studies and examination of the validity of the SAT support the decisions of colleges and universities to use entrance examinations as but one indicator of college success. Such tests may help discriminate at a very broad level between students in the eligible and ineligible group. They have limited predictive power for fine-grained selection decisions. Thus, MIT and other institutions identify a pool of eligible applicants using college entrance examinations, grades, recommendations, and other indicators. They base decisions among those in the eligible pool on indicators other than college entrance examinations. This approach values high school rank in class and breadth of accomplishments. It paves the way for considering original work such as projects or performances.

Overall, aptitude test scores favor males, yet the validity of these scores for predicting college performance, for selecting individuals for scholarships, for placement, and for inferring mathematical aptitude is in question. As a result, many colleges and universities across the United States are modifying admissions practices to adjust for the gender gap on college admissions mathematics examinations. This gender gap does not appear to have predictive validity for success in undergraduate programs. Additionally, these tests convey a misleading picture of the nature of mathematics. College entrance examinations are biased predictors of college grades and should not be used alone.

Social context of mathematics

How does the context of mathematical learning contribute to the pattern of grades and scores reported here? Theoretical and empirical work suggests [32, 48, 83, 135] that women are less confident than males about mathematics and that both males and females stereotype mathematics as a male domain [48]. This difference in confidence and in perception of mathematics as a male domain may contribute to different learning practices. Many examples of the comments reflecting low confidence appear above in the comments of students who switched out of mathematics. Learners who lack confidence in their ability in mathematics rely more on opinions of peers and professors, follow instructions provided by instructors, check their work numerous times, fail to contribute to class discussions, and have low expectations about their abilities. In general, unconfident learners take test scores and grades more seriously, and interpret the behavior of peers and faculty more negatively. Dweck [31] describes females as believing they are guilty (likely to fail) until proven innocent.

Most people expect males to be more successful than females, probably based on the historical predominance of males in mathematics. Consistent with the normative view that women cannot do mathematics, balance their checkbooks, or handle investments, score differences may reinforce stereotypes and subtly discourage females. Rather than recognizing the largely overlapping distributions of performance of males and females on tests or grades, most expect males in general to outperform females in general [71].

Females expect that they will earn lower scores and grades than they actually earn. Males expect to earn higher grades and scores than they actually earn. Females report that faculty reinforce these stereotypes (see Figure 10).

Many feel that they are less encouraged to continue in mathematics than male students. Thus females report lower expectations about their success in mathematics on average than males and see this as consistent with faculty perceptions. Males are more likely than females to view themselves as successful and capable in mathematics. How do these different self-perceptions contribute to course and test performance of males and females?

We have argued that students who are concerned about their success will develop better study habits and engage in more preparation for class. In classes where teachers encourage good study habits, females report using them more than males [81]. Concerned students may gain a more cohesive and integrated

There weren't many women in our classes, but there were a number of A.B. women in freshman chemistry and physics and organic chemistry. In those large lecture halls, the ratios didn't seem all that terrible. But math classes were another story. Only once did I find myself to be the only woman in a class. . . . It was also the only class I took where the professor obviously disapproved of a woman being there. Maybe he wouldn't have acted the same [way] if there had been other women in the class. How many times I heard him ask if I understood what was being discussed."

—Princeton student [91, p. 64]

I found my introductory course in math to be very discouraging after advanced math in high school my junior year, which I excelled in. My intro course dampened my passion for math."

—Wellesley student [95, p. 37]

She's teaching a whole different way from the way I learned it. I'm used to taking shortcuts and in her class you cannot take a shortcut. You have to go from one step to the next. If you miss a step, the problem is wrong even if you come up with the right answer."

—Student at southern college [45, p. 169]

I definitely got frustrated. Freshman year was particularly bad, taking PHY 105 and MATH 203/204. I started in MATH 217 and dropped right down. That was a real confidence-buster. After that, I was used to not understanding things right away. Later, I found I could understand in time. The problem wasn't that the courses were too hard, but the sheer volume was [overwhelming]. I got through it, and I have to say that this year has been better. It's finally starting to pay off."

—Princeton student [91, p. 38]

One of my calculus professors; I found him very annoying. When he would say anything and a girl would ask him a question, he would say girls don't have to know because girls don't have to study math. He meant girls in general, not me in particular."

—Student at large northeastern university [79, p. 8]

I had extremely good math teachers in my high school. They could always explain things five different ways, and one of the five usually clicked with someone. I came here and was really discouraged because, although the people teaching math had Ph.D.'s, and my teachers probably didn't have more than a master's, and a teaching certificate, I thought they taught a whole lot better . . . Actually, I'm surprised this college retains as many students as they do."

—Student at southwestern university [44, p. 109]

FIGURE 10.   Comments of female students about mathematics instruction in undergraduate programs.

view of the subject.

In classes where mathematics is routinized, students are encouraged to learn algorithms, and tests reinforce this approach. Concerned students may have powerful algorithms but fewer alternative methods and less experience developing personalized algorithms [62]. In these classes, students who conform to the requirements may become bored and leave mathematics.

On mathematics tests, concerned students may rely on the algorithms that were taught and engage in extra checking of answers. This concern could penalize students on timed tests.

For example, to account for gender differences on the SAT-M, it is necessary to explain why females get about three more items wrong, on average, than males. By taking extra time to check answers, concerned students may not have time for the later items that they could perform correctly. By checking easy items, or skipping these, concerned students may allocate time inefficiently. Data from National Assessment of Education Progress items that offer an "I don't know" option shows that females use this response far more than males [75]. On science tests females also skip easy items more frequently [78].

In summary, the differential effect of the social context of mathematics on males and females could account for observed differences in grades and entrance examination scores. The social context combined with the goals of introductory college courses to "weed out" students could account for more female than male switching. Additionally, courses that emphasize algorithms and routine problems could discourage the most talented males and females.

Conclusions and recommendations

To retain the best students in mathematics, courses must improve. Far too many talented students leave college mathematics and many more are women than men. Indeed, the best predictor of finishing a Ph.D. in mathematics is the gender of an individual. Males are far more likely to earn a Ph.D. than females.

Undergraduate courses must require transformation because they fail to communicate the nature of mathematics and may attract just the wrong students to mathematics. Females, who earn higher grades, may be the most influenced. By learning the lessons in these courses successful students may discover they prefer other fields.

To improve mathematics instruction for all students these findings point towards communicating to students an authentic view of the field of mathematics. Current courses and college entrance examinations tend to reinforce stereotypes about the field and its participants. They probably discourage the wrong students. Adding projects that require deeper understanding will communicate a more robust and realistic view of mathematics.

This recommendation coincides with evidence that a few small liberal arts colleges produce significant numbers of female majors, graduate students, and Ph.D.'s in mathematics (see Tables 4–6). Such colleges offer close relationships between students and faculty, opportunities for undergraduate research experience, and a preponderance of female faculty members. Most students participate in discussions with faculty, increasing the likelihood that students gain deep understanding of the nature of mathematics [107]. These schools engage a more diverse group of students in the discourse of mathematics.

These findings also imply that faculty should emphasize nurturing rather than filtering. Students in mathematics are excellent at filtering themselves out of mathematics. Nurturing is necessary to sustain the best and brightest students. Programs that nurture students are showing success at St. Olaf and Potsdam [25, 38, 94, 98, 112, 116]. Especially when coming from successful high school programs to competitive college experiences, the best students are the most likely to be deterred by traditional courses. If "weeder" courses require only superficial understanding of mathematics, the best students may switch to other fields.

This recommendation has been implemented at the Massachusetts Institute of Technology where students get no first year grades, and learn that in first early courses it will be difficult for them to determine their relative standing compared to other students because of the differential previous experiences that students have. Many institutions help students understand opportunities in mathematics. UCLA nurtures students by providing a full time undergraduate advisor dedicated to helping students by, for example, distributing accurate information about jobs for mathematics majors and forming supportive, collaborative groups [51, 127].

These findings also suggest the advantage of monitoring behavior and examining activities that may subtly reinforce stereotypes. Faculty techniques that are successful for male students may not work for females. Thus, courses that filter rather than nurture may filter out fewer potentially successful males and a larger proportion of potentially successful females as indicated in the many quotes from students reported earlier. For example, discourse that emphasizes heated arguments may silence more females than males.

Faculty who display indifference towards all students may have a far greater effect on females than males. Males may simply assume that faculty are disinterested in students altogether, where females may take this more personally because of (a) their greater uncertainty about their ability to succeed in the field [89], and (b) their awareness of the general societal view that mathematics is a male domain.

Faculty frequently complain that they cannot monitor their behavior and that students should understand that they are well-meaning. Our examination suggests that this view is short-sighted. All members of society need to pay attention to the subtle messages they communicate. Numerous books illustrate how these messages can inhibit effective collaboration (e.g., [53, 56, 122].

Reinforcing and perpetuating stereotypes about women in mathematics works to discourage females. Many aspects of the culture of mathematics including the discourse style and patterns of interaction are developed within a same-sex group. They often silence females when practiced in a mixed-sex setting.

These findings also underscore the need to help the best students recognize their potential. Most students benefit from feedback and guidance concerning their strengths. Female students who have to overcome expectations that their gender does not succeed in mathematics need specific feedback to help them make sense of their accomplishments. For example, many female students interpret a grade of A– negatively.

A corollary of this recommendation is that instruction is most successful when it helps all students become autonomous learners and lifelong problem solvers, when it helps all students develop internal standards for success rather than relying primarily on external indicators of success, and when it helps all students interact in a community of mutual respect rather than isolating some students and stereotyping others. Since mathematics has traditionally been a male domain and since the predominant perception among men is that it continues to be a male domain, it is particularly important to reach out to underrepresented groups, to encourage the organizations that support these individuals, and to engage in community-building that cuts across groups [6].

REFERENCES

1. Adelman, C., *Women at Thirtysomething: Paradoxes of Attainment*, U.S. Department of Education, Office of Educational Research and Improvement, Washington, DC, 1991.
2. Agogino, A. M. and Linn, M. C., *Retaining female engineering students: Will early design experiences help? [Viewpoint Editorial]*, NSF Directions **5(2)** (1992), 8–9.
3. Astin, A. and Astin, H., *Undergraduate science education: The impact of different college environments on the educational pipeline in the sciences*, Higher Education Institute, Los Angeles, CA, 1993.
4. Blank, R. K. and Grubel, D., *State Indicators of Science and Mathematics Education 1993* (1993), Council of Chief State School Officers, Washington, DC.
5. Boli, J., Allen, M. L., and Payne, A., *High-ability women and men in undergraduate mathematics and chemistry courses*, American Educational Research Journal **22** (1985), 605–626.
6. Bonsangue, M. V., *An efficacy study of the calculus workshop model*, CBMS Issues in Mathematics Education: Research in Collegiate Mathematics Education. 1 **4** (1994), 117–137.
7. Bridgeman, B. and Wendler, C., *Gender differences in predictors of college mathematics performance and in college mathematics courses.*, Journal of Educational Psychology **83(2)** (1991), 275–284.
8. Brown, M., Shure, P., Megginson, B., Shaw, D., and Black, B., *Math 115 Calculus: Instructor's Guide*, University of Michigan; Mathematics Department, Ann Arbor, Michigan, Fall 1994.
9. Bruer, J. T., *Schools for Thought: A Science of Learning in the Classroom*, MIT Press, Cambridge, MA, 1993.
10. Buerk, D., *Carolyn Werbel's journal: Voicing the struggle to make meaning of mathematics*, Tech. Rep. No. 160 (Working paper) (1986), Wellesley College, Center for Research on Women, Wellesley, MA.
11. Burton, N. W., *How have changes in the SAT affected women's mathematics performance?*, Paper presented at the American Educational Research Association Annual Meeting, San Francisco, CA (April 1995).
12. Chipman, S. F., Brush, L. R., and Wilson, D. M. (Ed.), *Women and Mathematics: Balancing the Equation*, Lawrence Erlbaum Associates, Hillsdale, NJ, 1985.
13. Chronicle of Higher Education, *Seven institutions report they benefited from dropping SAT's* (1987).
14. Cipra, B., *At state schools, calculus reform goes mainstream*, Science **260** (1993), 484–485.
15. Cipra, B., *The bumpy road to reform*, UME Trends **6(6)** (1995), 16, 19.
16. Clancy, M. J. and Linn, M. C., *Designing Pascal solutions: A case study approach*, Principles of Computer Science, 1st ed. (A. V. Aho and J. D. Ullman, eds.), W. H. Freeman and Company, New York, NY, 1992.

17. Clark, M. J. and Grandy, J., *Sex differences in the academic performance of scholastic aptitude test takers*, Tech. Rep. No. 84-8, ETS RR No. 84-43 (College Board Report) (1984), College Entrance Examination Board, New York.

18. Clarke, D., *Using assessment to renegotiate the didactic contract*, paper presented at the Annual meeting of the American Education Research Association, San Francisco (April 18–22 1995).

19. Cohen, E. G., *Restructuring the classroom: Conditions for productive small groups*, Review of Educational Research **64(1)** (1994), 1–35.

20. Cohen, M., Knobel, A., Kurtz, D., and Pengelly, D. J., *Making calculus students think with research projects*, Mathematical Thinking and Problem Solving (A. Schoenfeld, ed.), Erlbaum, Hillsdale, NJ, 1994, pp. 193–208.

21. Collins, A., Brown, J. S., and Holum, A., *Cognitive apprenticeship: Making thinking visible*, American Educator **15(3)** (1991), 6–11, 38–39.

22. Conference Board of the Mathematical Sciences, *Undergraduate Survey*, 1990.

23. Connor, K. and Vargyas, E., *The legal implications of gender bias in standardized testing*, Berkeley Women's Law Journal **7** (1992).

24. Culotta, E., *The calculus of education reform (New teaching methods for college calculus)*, Science **255(5048)** (1992), 1060.

25. Datta, D. K., *Math education at its best: The Potsdam model*, Rhode Island Desktop Enterprises, Kingston, RI, 1993.

26. Davis, E. A., Linn, M. C., Mann, L. M., and Clancy, M. J., *Mind your Ps and Qs: Using parentheses and quotes in LISP*, paper presented at the Fifth Workshop on Empirical Studies of Programmers, Palo Alto, CA, Empirical Studies of Programmers: Fifth Workshop (C. R. Cook, J. C. Scholtz, and J. C. Spohrer, eds.), Ablex, Norwood, NJ, 1993, pp. 62–85.

27. de Lange, J., *Mathematics, Insight and Meaning*, Vakgroep Onderzoek Wiskundeonderwijs en Onderwijscomputercentrum, Utrecht, The Netherlands, 1987.

28. DeBoer, G., *A study of gender effects in science and mathematics course-taking behavior among students who graduated from college in the late 70's*, Journal of Research in Science Teaching **21** (1984), 95–103.

29. Douglas, R. G., *Toward a lean and lively calculus: Report of the conference/workshop to develop alternative curriculum and teaching methods for calculus at the college level*, MAA Notes, No. 6, Mathematical Association of America, Washington, DC, 1986.

30. Dowd, M., *The bitch factor*, Working Woman **16(6)** (1991), 78–80, 98.

31. Dweck, C., *Motivational processes affecting learning*, American Psychologist **41** (1986), 1040–1048.

32. Eccles, J. S., Wigfield, A., Flanagan, C. A., Miller, C., Reuman, D. A., and Yee, D., *Self-concepts, domain values, and self-esteem: Relations and changes at early adolescence*, Journal of Personality **57(2)** (1989), 283–309.

33. Elliot, R. and Strenta, A. C., *Effects of improving the reliability of the GPA on prediction generally and on comparative predictions for gender and race particularly*, Journal of Educational Measurement **25(4)** (1988), 333–347.

34. Ethington, C. A., *Gender differences in mathematics: An international perspective*, Journal of Research in Mathematics Education **21** (1990), 74–80.

35. Frazier-Kouassi, S. et al., *Women in Mathematics and Physics: Inhibitors and Enhancers*, University of Michigan, Ann Arbor, MI, 1992.

36. Friedman, L., *Mathematics and the gender gap: A meta-analysis of recent studies on sex differences in mathematical tasks*, Review of Educational Research **59(2)** (1989), 185–213.

37. Gallagher, A. M., *Sex Differences in problem-solving strategies used by high scoring examinees on the SAT-M*, College Entrance Examination Board, New York (to appear).

38. Gilmer, G. and Williams, S., *An interview with Clarence Stephens*, UME Trends **2(1)** (1989), 1, 4, 7.

39. Gipps, C. and Murphy, P., *A fair test? Assessment, Achievement, and Equality*, Open University Press, Philadelphia, PA, 1994.

40. Grandy, J., *Ten-year trends in SAT scores and other characteristics of high school seniors taking the SAT and planning to study mathematics, science, or engineering*, Educational Testing Service, Princeton, NJ, 1987.

41. Gross, B., *Bitch*, Salmagundi **103** (1994), 146–157.

42. Gross, S., *Participation and performance of women and minorities in mathematics* (1988), Department of Educational Accountability, Montgomery Public Schools, Maryland.

43. Hanna, G., *Mathematics achievement of girls and boys in grade eight: Results from twenty countries*, Educational Studies in Mathematics **20** (1989), 225–232.

44. Hewitt, N. M. and Seymour, E., *Factors contributing to high attrition rates among science, mathematics, and engineering undergraduate majors* (1991), Alfred P. Sloan Foundation, New York.

45. Holland, D. and Eisenhart, M., *Educated in Romance: Women, Achievement, and College Culture* (1990), University of Chicago Press, Chicago, IL.

46. Hughes, R., *Calculus reform and women undergraduates.*, L. Steen (Ed.) Calculus for a New Century: A Pump, Not a Filter, MAA Notes, No. 8, Mathematical Association of America, Washington, DC, 1988, pp. 125–129.

47. Humphreys, Sheila M. and Freeland, Robert, *Retention in engineering: A study of freshman cohorts*, University of California at Berkeley, College of Engineering, Berkeley, CA, 1992.

48. Hyde, J. S., Fennema, E., and Lamon, S. J., *Gender differences in mathematics performance*, Psychological Bulletin **107** (1990), 139–155.

49. Jackson, A., *Perspectives on the underrepresentation of minorities in mathematics: An interview with James C. Turner, Jr.*, Notices of the American Mathematical Society **41(5)** (1994), 448–450.

50. Johnson, E. S., *College women's performance in a math-science curriculum: A case study*, College and University **68(2)** (1993), 74–78.

51. Johnson, L., *What can I do with a degree in math? Your guide to degree opportunities*, University of California at Los Angeles Mathematics Department, Los Angeles, CA, 1992.

52. Kanarek, E., *Gender differences in academic performance and their relationship to the use of the SAT in admissions.*, paper presented at the Annual meeting of the Northeast Association for Institutional Research, Providence, RI (October 1988).

53. Katz, M. and Vieland, V., *Get smart!: What you should know (but won't learn in class) about sexual harassment and sex discrimination*, The Feminist Press at the City University of New York, New York, New York, 1993.

54. Keith, S., *How do students feel about calculus reform, and how can we tell?*, UME Trends **6(6)** (1995), 6, 31.

55. Keller, E. F., *Reflections on Gender and Science*, Yale University Press, New Haven, CT, 1985.

56. Kenschaft, P. C. (Ed.)., *Winning women into mathematics*, Mathematical Association of America, Washington, DC, 1991.

57. Kimball, M. M., *A new perspective on women's math achievement*, Psychological Bulletin **105(2)** (1989), 198–214.

58. Koehler, M. S., *Classrooms, teachers, and gender differences in mathematics*, Mathematics and Gender (E. Fennema and G. Leder, eds.), Teachers College Press, New York, 1990, pp. 128–148.

59. Lavin, D. E., *The Prediction of Academic Performance*, Russell Sage Foundation, New York, 1965.

60. Leder, G., *Using research productively*, Mathematics Without Limits (C. Beesey and D. Rasmussen, eds.), Mathematical Association of Victoria, Brunswick, Australia, 1994, pp. 67–72.

61. Leder, G. C., *Teacher/student interactions in the classroom: A different perspective*, E. Fennema and G. Leder (Ed.), Mathematics and gender, Teachers College Press, New York, 1990, pp. 149–168.

62. Leinhardt, G., Seewald, A. M., and Engel, M., *Learning what's taught: Sex differences in instruction*, Journal of Educational Psychology **71(4)** (1979), 432-439.

63. Leonard, D. and Jiang, J., *Gender bias in the college predictions of the SAT*, paper presented at the 1995 Annual Meeting of the AERA, San Francisco, CA.

64. Leonard, D. K. and Jiang, J., *Gender bias in the SAT's predictions of undergraduate performance at Berkeley*, paper presented at the 1994 Annual Meeting of the AERA, New Orleans, LA.

65. Levin, J. and Wyckoff, J., *Effective advertising: Identifying students most likely to persist and succeed in engineering*, Engineering Education **78** (1988), 178–182.

66. Lewis, C., *Why and how to learn why: Analysis-based generalization of procedures*, Cognitive Science **12** (1988), 211–256.

67. Linn, M. C., *Teaching children to control variables: Some investigations using free choice experiences*, Toward a Theory of Psychological Development Within the Piagetian Framework (S. Modgil and C. Modgil, eds.), National Foundation for Educational Research Publishing Company, Windsor, Berkshire, England, 1980.

68. Linn, M. C., *Science*, Cognition and Instruction (R. Dillon and R. J. Sternberg, eds.), Academic Press, New York, 1986, pp. 155–204.

69. Linn, M. C., *Gender differences in educational achievement*, Sex equity in educational opportunity, achievement, and testing (Proceedings of 1991 Educational Testing Service Invitational Conference) (J. Pfleiderer, ed.), Educational Testing Service, Princeton, NJ, 1992, pp. 11–50.

70. Linn, M. C., *Gender and school science*, The International Encyclopedia of Education (T. Husén and T. N. Postlethwaite, eds.), vol. 4, 2nd ed., Pergamon Press, New York, 1994, pp. 2436–2440.

71. Linn, M. C., *The tyranny of the mean: Gender and expectations*, Notices of the American Mathematical Society **41(7)** (1994), 766–769.

72. Linn, M. C., *Designing computer learning environments for engineering and computer science: The scaffolded knowledge integration framework*, Journal of Science Education and Technology **4(2)** (1995), 103–126.

73. Linn, M. C. and Burbules, N. C., *Construction of knowledge and group learning*, The Practice of Constructivism in Science Education (K. Tobin, ed.), American Association for the Advancement of Science, Washington, DC, 1993, pp. 91–119.

74. Linn, M. C. and Clancy, M. J., *Can experts' explanations help students develop program design skills?*, International Journal of Man-Machine Studies **36(4)** (1992), 511–551.

75. Linn, M. C., de Benedictis, T., Delucchi, K., Harris, A., and Stage, E., *Gender differences in National Assessment of Educational Progress science items: What does "I don't know" really mean?*, Journal of Research in Science Teaching **24(3)** (1987), 267–278.

76. Linn, M. C. and Hyde, J. S., *Gender, mathematics, and science*, Educational Researcher **18(8)** (1989), 17–19, 22–27.

77. Linn, M. C. and Petersen, A. C., *Emergence and characterization of sex differences in spatial ability: A meta-analysis*, Child Development **56** (1985), 1479–1498.

78. Lockheed, M. E., *Sex equity in classroom organization and climate*, Handbook for Achieving Sex Equity Through Education (S. S. Klein, ed.), The Johns Hopkins University Press, Baltimore, MD, 1985, pp. 189-217.

79. Lyons-Lepke, E. M., *Choosing majors and careers: An investigation of women persisters and defectors in mathematics and science majors in college*, (unpublished doctoral dissertation) (1986), Rutgers University,, New Brunswick, NJ.

80. Mann, L., *The relationship between study techniques and success of women and men*, paper presented at the Taking the Lead: Balancing the Educational Equation, "Symposium, Gender, Mathematics, Science in College" (October 24, 1992), Mills College, Oakland, California.

81. Mann, L. M., Linn, M. C., and Clancy, M. J., *Evaluating an evaluation modeler*, (Technical report) (1992), University of California, Hypermedia Case Studies in Computer Science, Berkeley, CA.

82. McDade, L. A., *"Knowing the right stuff": Attrition, gender and scientific literacy*, Anthropology and Education Quarterly **19** (1988), 93–114.

83. Meece, J. L., Parsons, J. E., Kaczala, C. M., and Goff, S. B., *Sex differences in math achievement: Toward a model of academic choice*, Psychological Bulletin **91(2)** (1982),

324–348.

84. Miller, R. R., *Letter to the editor*, UME Trends **2(4)** (1990), 7.

85. Murphy, R. J. L., *Sex differences in objective test performance*, British Journal of Educational Psychology **52** (1982), 213–219.

86. National Center for Educational Statistics, *Data compendium for the NAEP 1992 mathematics assessment of the nation and the states.* (1993), U.S. Department of Education, Office of Educational Research and Improvement, Washington, DC.

87. National Center for Educational Statistics, *A preliminary report of the national estimates from the National Assessment of Education Progress 1992 mathematics assessment* (1993), U.S. Department of Education, Washington, DC.

88. National Science Foundation, *Women and minorities in science and engineering* (1990), National Science Foundation, Washington, DC.

89. Nerad, M., *Using time, money, and human resources efficiently and effectively in the case of women graduate students*, paper presented at the Conference on Science and Engineering Programs: On Target for Women?, Irvine, CA (Nov 4–5 1991).

90. Nevitte, N., Gibbins, R., and Codding, P. W., *The career goals of female science students in Canada*, The Canadian Journal of Higher Education **18(1)** (1988), 31–48.

91. Ng, Y. and Rexford, J., *She's an engineer?: Princeton alumnae reflect* (1993), Princeton University, Princeton, NJ.

92. Office of Institutional Research/Office of Student Research, *Berkeley Campus Statistics Fall 1992* (1992), University of California, Berkeley, Berkeley, CA.

93. Petersen, A. C. and Dubas, J. S., *Strategies for achieving gender equity in postsecondary education*, Sex Equity in Educational Opportunity, Achievement, and Testing, Educational Testing Service, Princeton, NJ, 1992.

94. Poland, J., *A modern fairy tale?*, American Mathematical Monthly **94(3)** (1987), 291–295.

95. Rayman, P. and Brett, B., *Pathways for women in the sciences* (1993), Wellesley Center for Research on Women, Wellesley, MA.

96. Reynolds, C. R., *The problem of bias in psychological assessment*, The Handbook of School Psychology (C. R. Reynolds and T. B. Gutkin, eds.), John Wiley and Sons, New York, 1982, pp. 178–208.

97. Robinson, A. and Katzman, J., *Cracking the System* (1986), Villard Books, New York.

98. Rogers, P., *Thoughts on power and pedagogy*, Gender and Mathematics: An International Perspective (L. Burton, ed.), Cassell, London, 1992, pp. 38–45.

99. Rosser, P., *The SAT Gender Gap: Identifying the Causes* (1989), Center for Women Policy Studies, Washington, DC.

100. Sadker, M. and Sadker, D., *Sexism in the classroom: From grade school to graduate school*, Phi Delta Kappan (March 1986).

101. Sadker, M. and Sadker, D., *Failing at Fairness: How America's Schools Cheat Girls*, Maxwell Macmillan International, New York, 1994.

102. Sandler, B. and Hall, R., *The classroom climate: A chilly one for women?* (1982), Association of American Colleagues, Project on the Status and Education of Women, Washington, DC.

103. Schoenfeld, A., *Ideas in the air: Speculations on small group learning, environmental and cultural influences on cognition, and epistemology*, Berkeley Cognitive Science Report (Tech. Rep. No. 52) (1988), Institute of Cognitive Studies, Berkeley, CA.

104. Schoenfeld, A., *Reflections on doing and teaching mathematics*, Mathematical Thinking and Problem Solving (A. Schoenfeld, ed.), Erlbaum, Hillsdale, NJ, 1994, pp. 53–70.

105. Schoenfeld, A., *A brief biography of calculus reform*, UME Trends **6(6)** (1995), 3–5.

106. Selden, A. and Selden, J., *A math ed miscellany from Vancouver*, Undergraduate Mathematics Education Trends **5(5)** (1993), 1, 5, 8.

107. Senechal, L., *Models for Undergraduate Research in Mathematics*, MAA Notes, No. 18, Mathematical Association of America, Washington, DC, 1991.

108. Seymour, E. and Hewitt, N., *Talking about leaving: Factors contributing to high attrition rates among science, mathematics, and engineering undergraduate majors (Report on*

an ethnographic inquiry at seven institutions) (1994), Alfred P. Sloan Foundation, New York.

109. Sheehan, K. R., *The relationship of gender bias and standardized tests to the mathematics competency of university men and women*, (unpublished doctoral dissertation, American University) (1989).

110. Simpson, J. C., *Middlebury is joining other colleges dropping SAT admission requirement*, Wall Street Journal (April 15, 1987), 34, 38.

111. Songer, N. B., *Learning science with a child-focused resource: A case study of Kids as Global Scientists*, Proceedings of the Fifteenth Annual Meeting of the Cognitive Science Society, Lawrence Erlbaum Associates, Hillsdale, NJ, 1993, pp. 935–940.

112. Spencer, A., *On attracting and retaining mathematics majors—Don't cancel the human factor*, Notices of the American Mathematical Society **42(8)** (1995), 859–862.

113. Spertus, E., *Why are there so few female computer scientists?*, Tech. Rep. No. AITR 1315 (1991), Massachusetts Institute of Technology, Artificial Intelligence Laboratory, Cambridge, MA.

114. State Department of New York, *Sharif v. New York State Education Department*, (Legal proceedings, 88 Civ. 8435) (1989), 14.

115. Steen, L. A. (Ed.), *Calculus for a New Century: A Pump, Not a Filter*, MAA Notes, No. 8, Mathematical Association of America, Washington, DC, 1988.

116. Steen, L. A. (Ed.), *Reshaping College Mathematics: A project of the Committee on the Undergraduate Program in Mathematics*, MAA Notes, No. 13, Mathematical Association of America, Washington, DC, 1989.

117. Stobart, G., Elwood, J., and Quinlan, M., *Gender bias in examinations: How equal are the opportunities?*, British Educational Research Journal **18(3)** (1992), 261–276.

118. Strauss, S. M. and Subotnik, R. F., *Gender differences in behavior and achievement: A true experiment involving random assignment to single sex and coeducational advanced placement (BC) calculus classes*, Association for Women in Mathematics Newsletter **24(3)** (1994), 12–25.

119. Stricker, L. J., Rock, D., and Burton, N. W., *Sex differences in SAT predictions of college grades*, (College Board Report No. 91-2) (1991), College Entrance Examination Board, New York.

120. Struik, R. R. and Flexer, R., *Self-paced calculus: A preliminary evaluation*, American Mathematical Monthly **84(2)** (1977), 129–134.

121. Struik, R. R. and Flexer, R., *Sex differences in mathematical achievement: Adding data to the debate*, International Journal of Women's Studies **7(4)** (1984), 336–342.

122. Tannen, D., *Talking from 9 to 5: How Women's and Men's Conversational Styles Affect Who Gets Heard, Who Gets Credit, and What Gets Done at Work*, William Morrow and Company, Inc., New York, New York, 1994.

123. Thorndike, R. L., *The concepts of over- and under-achievement*, Bureau of Publications, Teachers College, Columbia University, New York, 1963.

124. Tobias, S., *They're Not Dumb, They're Different: Stalking the Second Tier*, Research Corporation, Tucson, AZ, 1990.

125. Tucker, A. C. and Leitzel, J. R. C., *Assessing Calculus Reform Efforts: A Report to the Community*, Mathematical Association of America, Washington DC, 1995.

126. Tucker, T. W., *Priming the Calculus Pump: Innovations and Resources*, MAA Notes, No. 17, Mathematical Association of America, Washington, DC, 1990.

127. University of California at Los Angeles Mathematics Department, *UCLA Department of Mathematics 1993–94 Undergraduate Handbook*, 1993.

128. University of Michigan, *The new calculus at the University of Michigan progress report* (1993).

129. US Department of Education/National Center for Education Statistics, *Completions Survey/Fall Enrollment Survey*, (CASPAR Database System) (1994), National Science Foundation, Washington, DC.

130. Wagner, M. E. and Strabel, E., *Homogeneous grouping as a means of improving the prediction of academic performance*, Journal of Educational Psychology **19** (1935), 426–46.

131. Wainer, H. and Steinberg, L. S., *Sex differences in performance on the mathematics section of the Scholastic Aptitude Test: A bidirectional validity study*, Harvard Educational Review **62(3)** (1992), 323–336.

132. Webb, N. M., *Peer interaction and learning in small groups*, International Journal of Educational Research **13(1)** (1989), 21–39.

133. Wellesley College Center for Research on Women, *How Schools Shortchange Girls*, American Association of University Women Educational Foundation, Washington, DC, 1992.

134. White, P. E., *Women and Minorities in Science and Engineering: An Update*, National Science Foundation, Washington, DC, 1992.

135. Wigfield, A., Eccles, J. S., MacIver, D., and Reuman, D. A., *Transitions during early adolescence: Changes in children's domain-specific self-perceptions and general self-esteem across the transition to junior high school*, Developmental Psychology **27(4)** (1991), 552–565.

136. Willingham, W. W., Lewis, C., Morgan, R., and Ramist, L., *Predicting College Grades: An Analysis of Institutional Trends Over Two Decades*, Educational Testing Service, Princeton, NJ, 1990.

137. Wise, L. L., *Project TALENT: Mathematics course participation in the 1960s and its career consequences*, Women and Mathematics: Balancing the Equation (S. F. Chipman, L. R. Brush, and D. M. Wilson, eds.), Erlbaum, Hillsdale, NJ, 1985, pp. 25–58.

138. Young, J. W., *Gender bias in predicting college academic performance: A new approach using item response theory*, Journal of Educational Measurement **28(1)** (1991), 37–47.

139. Zaslavsky, C., *Fear of Math: How to Get Over It and Get On With Your Life*, Rutgers University Press, New Brunswick, NJ, 1994.

GRADUATE SCHOOL OF EDUCATION, UNIVERSITY OF CALIFORNIA, BERKELEY

Appendix: Summary of sociological and ethnographic studies cited

Study	Institution	Methods and participants
Lyons-Lepke (1986, [79])	a large selective northwestern university	survey and individual interviews: 50 female undergraduates interested in majoring in math or science, SAT-M over 500.
McDade (1988, [82])	"State University" has entering class of about 4,000	interviews: 9 female chemistry switchers, 8 female mathematics switchers, 8 male chemistry switchers, 5 male mathematics switchers.*
Holland and Eisenhart (1990, [45])	2 institutions	individual interviews: 21 female undergraduates, high school grades of B+ or better, half expressed interest in math and science majors.
Tobias (1990, [124])	Harvard-Radcliffe	surveys: 300 female undergraduates, high "science aptitude" or interest in science. interviews: 80 of these students.
Hewitt and Seymour (1991, [44])	4 Southwestern institutions	individual interviews and focus groups: 149 undergraduates, majors or former majors in science, mathematics, and engineering.
Frazier-Kouassi et al. (1992, [35])	various undergraduate institutions	2 focus groups: 23 female graduate students in mathematics and physics at the University of Michigan.
Ng and Rexford (1993, [91])	Princeton	written reflections: 46 alumnae who received BAs in engineering, 1 alumna who received a BA in history, 2 female graduate students.
Rayman and Brett (1993, [95])	Wellesley	3 surveys: all undergraduates entering in 1991, the mean SAT-M score was 624.
	Howard	survey and 2 focus groups: 23 undergraduates.
	MIT	1 focus group of graduate students, 1 focus group of undergraduates.
Zaslavsky (1994, [139])	various, all levels of education	500 "math autobiographies"*
Seymour and Hewitt (1994, [108])	7 institutions	individual interviews and focus groups: 335 undergraduates, majors or former majors in science, mathematics, and engineering, SAT-M over 650.

*not "highly qualified"

CBMS Issues in Mathematics Education
Volume **6**, 1996

Analysis of Effectiveness of Supplemental Instruction (SI) Sessions for College Algebra, Calculus, and Statistics

SANDRA L. BURMEISTER, PATRICIA
ANN KENNEY, AND DORIS L. NICE

Introduction

Supplemental Instruction (SI) is a model of out-of-class learning assistance widely used in institutions of higher education. In this paper, we examine the data from 177 courses in mathematics for which SI support was given. The data indicate that SI support aids student success in these courses.

The SI model was developed by Deanna Martin of the University of Missouri-Kansas City (UM-KC) during the mid-1970's in an effort to address high attrition rates for students enrolled in difficult classes as part of their studies in the institution's six-year medical school program. Because of indications of success in the health science professional school, the SI program was extended to support many courses throughout the curriculum at UM-KC.

Since the 1970's, the model has received national recognition as an exemplary program from the U.S. Department of Education (in 1981 and again in 1992), and funding from the National Diffusion Network of the U.S.

Department of Education for replication in institutions of higher education (1984 to present). Currently, the model has been adopted by over 300 different institutions of higher education (Martin [13]).

Key components of the SI model include:

(1) SI study sessions are *attached to high risk courses* (those which have had a history of 30% or higher combined D's, F's, and withdrawals).

(2) SI sessions are non-remedial and *offered to all students* enrolled in the target courses. (In general, students enrolled in target courses attend SI sessions on a voluntary basis; however, both advisors and faculty frequently encourage students to attend SI sessions because of the students'

accomplishments or difficulties. The data we analyze in this paper do not give us information on which students attended sessions on a voluntary, semi-voluntary, or even mandatory basis.)

(3) SI sessions are *not tutoring sessions* but are supervised, small group, collaborative learning sessions.

(4) SI sessions are *conducted by a student or professional* who has been approved by the faculty member as competent in the course. Before the terms begin, all SI leaders complete a special training program designed to teach them how to conduct collaborative learning study sessions; and then leaders continue working with an SI supervisor throughout the term to enhance their group leadership skills.

(5) *SI supervisors and faculty monitor and evaluate* the effects of SI sessions on student success in the target courses. In any given term, as many as 200 different institutions of higher education send data to UM-KC as part of the dissemination project.

With the cooperation of the staff of the Center for Academic Development at UM-KC, we analyze in this paper the data from 45 separate institutions of higher education, institutions where SI study sessions were used for college algebra, calculus, and statistics classes.

Background

Martin and her colleagues designed the SI model to assist students in gaining mastery of course content and to address what they defined as students' developmental lack of competence in reading, reasoning, and study skills. Because the developers of SI conjectured about the underlying reasons for student failure to master course content, the program was originally grounded in the Piagetian notions of concrete and formal reasoning [16]. Martin and her colleagues suspected that many students continued to operate at the concrete (nonabstract) level of formal reasoning without progressing to formal operational (abstract) level (Blanc, DeBuhr, and Martin [1]). Subsequent analyses of the SI model found some potential links to the literature on metacognition (e.g., Kenney [9]; Kenney and Kallison [10]) in that the SI program emphasizes metacognitive ideas such as "knowing about what you know" and "regulating and monitoring what you are doing while you are doing it."

Current views of SI hold that during SI sessions students improve their learning and reasoning skills as well as learn course content necessary for success. Many researchers (e.g., Erlich and Kennedy [4]; Heiman and Slomianko [7]; Main [12]; Maxwell [14]; Pryor [17]) report success with academic support programs in which study skills such as notetaking, text-reading, and problem solving are integrated directly into the content for a particular academic course. This integration of study skills with course content remains an essential component of the SI model [1].

However, merely attaching sessions based on study skills to any academic course is not adequate. For students to attend additional study sessions on a voluntary basis, they must perceive the academic course to which the sessions are attached as challenging. In turn, the study sessions must then be effective in assisting students to master the content. The non-remedial image of an SI program is based first on carefully choosing the most appropriate high-risk course for the support, and secondly, training the SI leaders and supervising them to ensure that effective strategies are being modeled in the SI sessions (Martin, Arendale, and Associates [13]).

One common structure for SI sessions is based on cooperative learning. Today, there is a growing trend in higher education to implement collaborative learning techniques to enhance student learning. For example, Uri Treisman has incorporated cooperative learning in his mathematics study groups in the University of California–Berkeley Mathematics Workshop and the Emerging Scholars Program at the University of Texas at Austin (Garland [6]). In these workshops, students engage in challenging mathematical activities rather than in remedial exercises. There is some evidence that participation in the Treisman workshops has had an effect on the performance of minority students in entry-level mathematics courses such as calculus. For example, Treisman's project at UC–Berkeley improved the failure rate of minority students enrolled in calculus from a 60% failure rate before the workshop to a 4% failure rate after the workshop was available to students (Jackson [8]).

The collaborative learning model used in Treisman's program and other such programs is an important part of the model for SI sessions. Burmeister [2] describes collaborative learning in an SI session as follows:

"In an atmosphere which is safe for risk-taking, students generate ideas, evaluate and categorize these ideas, and teach one another as they work on common tasks. This 'cross teaching' (a phrase used by Moffit [15]) involves mutual contributions and mutual decisions regarding what to do and how to proceed."

One way that kind of collaborative learning takes place in SI sessions is in the form of paired problem solving, as suggested by Whimbey [18]. Students are matched in pairs and given problems which they select themselves or that are suggested by the SI leader. One student solves a problem while explaining every step to the partner, who often responds with probing questions such as "How do you justify that?" or "Are you sure about that?" The student solver must be able to justify the work a step at a time. When the problem is solved, the students reverse roles and begin a new problem. Techniques like this encourage students in SI groups to take risks, test their understanding, and work together toward success in mastering difficult course content. For further examples of collaborative learning techniques which have been used in SI sessions in mathematics see Burmeister [3].

Claims Investigated in this Analysis

Professionals on individual campuses support the SI model because they believe SI accomplishes the goal of student success with both long term and short term results. The short-term, in-class results are those which are considered in this paper. Through a review of the data from 45 institutions using SI to support student success in mathematics, we evaluated two claims of effectiveness made by proponents of the SI model:

Claim 1: Students participating in SI within the targeted high risk courses earn higher mean final course grades than students who do not participate in SI.

Claim 2: Students participating in SI within targeted high risk courses succeed at a higher rate (withdraw at a lower rate and receive a lower percentage of D or F final grades) than those who do not participate in SI (Martin [13]).

Selection of SI Mathematics Courses for Analysis

The staff of the Center for Academic Development at the University of Missouri-Kansas City supplied us with data for a variety of courses in mathematics for which SI sessions were offered. Our selection of college algebra, calculus, and statistics for this analysis is the result of our definition of these courses as both "high risk" and nonremedial. Our experience as SI supervisors has taught us that students are most motivated to attend SI sessions when the course material is difficult and when the course is a gatekeeper course for their major. All three of the courses we chose are mathematics-based courses required for a variety of majors in higher education.

We attempted to use data from courses not only labelled as college algebra, calculus, and statistics but also having similar course content (see Appendix). We believe that these courses are consistent over time and across institutions. For these courses then, students must demonstrate competency with a relatively consistent curriculum. All three courses are the first ones taken in that area of mathematics (all Calculus I, not Calculus II, or all Statistics I, not a second course in statistics); however, students may have had remedial or other courses in mathematics when they were enrolled in the course we analyzed, or they may have taken the course before since we do not have either piece of information regarding the students in the courses we analyzed.

The staff of the University of Missouri–Kansas City's Center for Academic Development supplied us with course evaluation sheets used by institutions to report the effects of SI sessions following the end of each term. The data included on these summary sheets give the final course grade or withdrawal status for each student enrolled in the course and separate students into SI participants and non-SI participants. The form also includes the average final course grades and the combined D, F, and withdrawal rates for both groups.

Analysis of Data for Supplemental Instruction in College Algebra, Calculus, and Statistics Courses

Our data for SI sessions for college algebra, calculus, and statistics courses was obtained from 45 different institutions. At these institutions, a total of 11,252 students were enrolled in a total of 177 classes for which SI support was offered—classes conducted during 16 terms. A total of 3,631 students (32% of those enrolled in these classes) attended SI sessions. Of the 2,824 students enrolled in college algebra, 1,130 (40%) attended SI sessions; of the 6,682 students enrolled in calculus, 1,875 (25%) attended SI sessions; and of the 1,746 students enrolled in statistics, 626 (26%) attended SI sessions. For both college algebra and calculus, the range of participation was 5% to 84%, and for statistics, the range of participation was 6% to 88%.

In the following sections, we present an analysis of comparisons between the SI and non-SI groups for each course. The groups were compared on average final course grades (using the standard conversion of A = 4.0, B = 3.0, . . . , F = 0.0) and on the frequencies of D and F grades and course withdrawals (W).

TABLE 1. Comparison of average final course grades for SI and non-SI groups.

	College Algebra 60 Classes on 21 Campuses	Calculus 78 Classes on 21 Campuses	Statistics 39 Classes on 14 Campuses
Average course grade: SI	2.21	2.28	2.49
Average course grade: Non-SI	1.98	1.83	2.32
t-value	$t = 2.514^*$	$t = 3.263^{**}$	$t = 2.394^{***}$

* significant at the .05 level, df = 2104   ** significant at the .01 level, df = 5740
*** significant at the .05 level, df = 1548

Average Final Course Grades

Table 1 contains the results of comparisons between the two groups for each course on final course grades, using two-tailed independent t-tests. For all three courses, the average course grade for the SI group was significantly different than the average course grade for the non-SI group. In terms of a more practical analysis of the average, the results for the college algebra and calculus courses have an encouraging interpretation. When the numerical averages were converted to their letter-grade equivalents, the college algebra and calculus students who attended SI sessions tended to earn grades above a "C" (2.21 for algebra; 2.28 for calculus), while the non-SI students earned grades below a "C" (1.98 for algebra; 1.83 for calculus). Moreover, in the case of programs which required students to earn at least a grade of "C" in order to advance to the next course (e.g., a "C" in Calculus I as a prerequisite for enrollment in Calculus II), the SI students have had a better chance of being able to continue in the normal sequence of mathematics courses, whereas the non-SI students might have been faced with repeating the course.

TABLE 2. Comparison of frequencies of
D's, F's, and W's for SI and non-SI groups.

College Algebra	SI		Non SI	
	Observed	Expected	Observed	Expected
D	153	104	167	216
F	107	107	227	226
W	185	234	538	490
$\chi^2 = 38.05$				

Calculus	SI		Non SI	
	Observed	Expected	Observed	Expected
D	259	200	570	628
F	183	197	630	616
W	183	228	757	713
$\chi^2 = 35.67$				

Statistics	SI		Non SI	
	Observed	Expected	Observed	Expected
D	81	53	97	125
F	59	62	151	148
W	31	56	156	131
$\chi^2 = 37.20$				

Frequencies of "D" and "F" Grades and Course Withdrawals ("W")

Table 2 contains the data relevant to the frequencies of D's, F's, and W's for the SI and non-SI groups for each of the three courses. The three separate Chi-square analyses for the data associated with college algebra ($\chi^2 = 38.05$), calculus ($\chi^2 = 35.67$), and statistics ($\chi^2 = 37.20$) all resulted in statistical significance at the .01 level. In all three cases, the SI students tended to earn more D's than expected; however, their rate of withdrawal from their respective courses was lower than expected. The non-SI participants showed a pattern of fewer D's than expected, but these students withdrew from the courses at a higher rate than the expected frequency. These results suggest that the persistence rate (as measured by completing the course with a "D" rather than withdrawing) for the SI students was higher than would be expected by chance. By at least completing their respective course, these students are likely to remain on track toward their goals of success (or eventual success) in the course.

Summary

The data presented here show a positive difference in grades for students who participate in SI sessions in college algebra, calculus, and statistics compared with those who do not participate. However, there are a number of questions which this kind of analysis leaves unanswered: How closely did each of the institutions for which there are data follow the SI model? How do the data compare between institutions which require attendance at SI sessions and those where attendance is voluntary? Are SI groups from campus to campus composed of students who can be predicted to do well or poorly? What is the correlation between increased attendance at SI sessions and level of success in the course? Is there a relationship between gender and participation in SI sessions? Do students who participate in SI sessions improve their general ways of studying for all classes? In particular, is there carryover in the next mathematics class? Do SI study groups become academic communities which continue on an academic or even social basis? And, finally, what effect does serving as an SI leader have on the SI leader's academic, interpersonal, and leadership skills?

References

1. Blanc, R.A., DeBuhr, L.E., and Martin, D.C., *Breaking the attrition cycle: the effects of supplemental instruction on undergraduate performance and attrition*, Journal of Higher Education **54(1)** (1983), 80–90.

2. Burmeister, S.L., *Collaborative learning in small groups*, Presentation at the National Association of Developmental Educators (NADE), Washington DC, March 19, 1993.

3. Burmeister, S.L., Carter, J.M., Hockenberger, L.R., Kenney, P.A., McLaren, A., and Nice, D.L., *SI Sessions in College Algebra and Calculus*, Supplemental Instruction: Increasing Student Achievement and Retention. New Directions for Teaching and Learning (D.C. Marten and D.R. Arendale, eds.), vol. 60, Jossey-Bass, San Francisco, 1994.

4. Erlich, H. and Kennedy, M., *Skills and content: coordinating the classroom*, Journal of Developmental and Remedial Education **6(3)** (1983), 24–27.

5. Flavell, J. H., *Metacognitive aspects of problem solving*, The Nature of Intelligence (L. Resnick, ed.), Lawrence Erlbaum, Hillsdale, NJ, 1976.

6. Garland, M., *The mathematics workshop model: an interview with Uri Treisman*, Journal of Developmental Education **16(3)** (1993).

7. Heiman, M. and Slomianko, J., *Learning to Learn: Some Questions and Answers*, Learning Associates, Cambridge, MA, 1987.

8. Jackson, A., *Minorities in mathematics: A focus on excellence, not remediation*, American Educator (Spring 1989), 22–27.

9. Kenney, P. A., *Effects of Supplemental Instruction (SI) on Student Performance in a College-level Mathematics Course*, (unpublished doctoral dissertation), Mathematics Education Division, Department of Curriculum and Instruction, The University of Texas at Austin, 1988.

10. Kenney, P. A. and Kallison, J. M., Jr., *Research studies on The effectiveness of supplemental instruction in mathematics*, Supplemental Instruction: Improving First-Year Student Success in high-Risk Courses, Jossey-Bass, San Francisco, 1994.

11. Kenney, P. A., *Effects of supplemental instruction on student performance in a college-level mathematics course*, Paper presented at the American Educational Research Association Conference, San Francisco (March, 1989).

12. Main, A., *Encouraging Effective Learning*, R. and R. Clark Publishers, Edinburgh, UK, 1980.

13. Martin, D. C., Arendale, D., and Associates, *Supplemental Instruction: Improving First-Year Student Success in High-Risk Courses*, National Resource Center for the Freshman Year Experience, University of South Carolina, Columbia, SC, 1992.

14. Maxwell, M., *Improving Student Learning Skills: A Comprehensive Guide to Successful Practices and Programs for Increasing the Performance of Underprepared Students*, Jossey-Bass, San Francisco, 1979.

15. Moffett, James, *Teaching the Universal Discourse* (1968), Houghton Mifflin, Boston.

16. Piaget,J. and Inhelder, B., *Growth of Logical Thinking*, Basic Books, New York, 1958.

17. Pryor, S.A., *The Relationship of Supplemental Instruction and Final Grades of Students Enrolled in High-risk Course*, (unpublished doctoral dissertation), Department of Education, Western Michigan University, 1989.

18. Whimbey, Arthur and Lochhead, Jack, *Problem Solving and Comprehension: A Short Course in Analytical Reasoning*, The Franklin Institute Press, Philadelphia, PA, 1980.

Appendix

The college algebra courses used in this study include the following common components: integer and rational exponents, polynomials and factoring, quadratic equations, inequalities and absolute value, graphing linear equations, polynomials and rational functions, conic sections, exponential and logarithmic functions, and systems of equations. The calculus courses used in this study include the following common components: limits, continuity, the derivative, differentiation, rate of change, the mean value theorem, the integral, and techniques and applications of integration.

Data on the effects of SI in college algebra and/or calculus were supplied by the following institutions:

Anne Arundle Community College, Arnold, MD
Azuza Pacific University, Azuza, CA
Baker University, Baldwin City, KS
Central Missouri State University, Warrensburg, MO
Colby Community College, Colby, KS
Cumberland University, Lebanon, TN
Deanza College, Cupertino, CA
Edmonds Community College, Lynnwood, WA
Fort Hays State University, Hays, KS
Genessee Community College, Batavia, NY
Glendale Community College, Glendale, CA
John Brown University, Siloam Springs, AR
Kean College, Union, NJ
Kingwood College, Kingwood, TX

Lawrence Technological university, Southfield, MI
Linn-Benton Community College, Albany, OR
Macomb Community College, Mount Clemens, MI
Milwaukee Area Technical College, Milwaukee, WI
Nashville State Technical Institute, Nashville, TN
North Carolina State University, Raleigh, NC
North Lake College, Irving, TX
Oakland University, Rochester, MI
Onondaga Community College, Syracuse, NY
Pennsylvania State University, University Park, PA
Salem State College, Salem, MA
Southern Arkansas University, Magnolia, AR
Trenton State College, Trenton, NJ
University of Idaho, Moscow, ID
University of Louisville, Louisville, KY
University of Maine, Orono, ME
University of Maine at Machias, Machias, ME
University of Missouri–St. Louis, St. Louis, MO
University of Nevada–Reno, Reno, NV
University of North Carolina at Charlotte, Charlotte, NC
University of Wisconsin–Parkside, Kenosha, WI
University of Utah, Salt Lake City, UT
Upsala College, East Orange, NJ
Wilmington College, Wilmington, OH

The statistics courses included the following common components: measures of central tendency, sampling theory (or techniques), hypothesis testing, frequency distributions, probability distributions, regression and correlation analysis.

Data on the effects of SI in statistics courses were supplied by the following institutions:

Baker University, Baldwin City, KS
DeAnza College, Cupertino, CA
Kutztown University, Kutztown, PA
Macomb Community College, Mount Clemens, MI
Nashville State Technical Institute, Nashville, TN
Point Loma Nazarene College, San Diego, CA
Salem State College, Salem, MA
Southwest State University, Marshall, MN
St. Edward's University, Austin, TX
Trenton State College, Trenton, NJ
University of Louisville, Louisville, KY

University of Minnesota–General College, Minneapolis, MN
University of North Carolina at Charlotte, Charlotte, NC
Washburn University, Topeka, KS
Western New Mexico University, Silver City, NM

HOBART AND WILLIAM SMITH COLLEGES, GENEVA, NEW YORK

LEARNING RESEARCH AND DEVELOPMENT CENTER, UNIVERSITY OF PITTSBURGH

UNIVERSITY OF WISCONSIN–PARKSIDE, KENOSHA, WISCONSIN

CBMS Issues in Mathematics Education
Volume **6**, 1996

A Comparative Study of a Computer-Based and a Standard College First-Year Calculus Course

KYUNGMEE PARK AND KENNETH J. TRAVERS

ABSTRACT. Calculus and *Mathematica* (C&M) is a computer laboratory calculus course that embraces many of the goals of the NSF calculus reform movement, including less emphasis on procedures and closed-form mechanical exercises and more attention to student participation, concept development and solving open-ended realistic problems. This article reports on a comparative study of outcomes (both cognitive and affective) for students in an experimental (C&M) and a standard first-year calculus course.

Generally, the findings from an achievement test, attitude survey, concept maps, and interviews were all favorable to the C&M students. The C&M group obtained a higher level of conceptual understanding than did the standard group without loss of computational proficiency. Furthermore, the C&M group's disposition toward mathematics and computers was far more positive than that of the standard group. This research employed assessment methods that go beyond the usual comparisons of test scores, thereby suggesting an alternative way to document the effects of projects designed to promote reform in undergraduate mathematics.

Introduction

Calculus and *Mathematica* (C&M) was first offered at the University of Illinois at Urbana–Champaign in the spring semester of 1989. Although several informal evaluations of C&M had been carried out prior to the present study, none was in depth. This research has two purposes: (1) to compare the mathematics achievement and attitudes of the C&M students with those of the students in the standard course; (2) to assess the effectiveness of the C&M course in attaining key goals of the calculus reform movement.

A Brief History of Calculus Reform

Two events, the Tulane Conference, "Toward a Lean and Lively Calculus," (1986) and the Washington, D.C., symposium, "Calculus for a New Century," (1987), signaled the beginning of calculus reform in the United States. The projects at Purdue and at Duke Universities, briefly described below, as well as Calculus and *Mathematica*, exemplify major themes of these reform initiatives.[1]

The themes of the calculus reform movement include: involving students in doing mathematics instead of lecturing at them; stressing conceptual understanding, rather than only computation; developing meaningful problem-solving abilities, not just "plug-and-chug"; exploring patterns and relationships, instead of just memorizing formulas; becoming engaged in open-ended, discovery-type problems, rather than doing routine, closed-form exercises; and approaching mathematics as a live exploratory subject, not merely as a description of past work (Small and College, [14]; see also NSF Program Guide, Calculus [11]).

Laboratory work in mathematics has become more frequent since the advent of the personal computer and the availability of appropriate numerical, graphic, and symbolic software. Nonetheless, many faculty members are reluctant to adopt a laboratory approach to calculus. One concern is that on many campuses adequate computer laboratory facilities are not yet available. Furthermore, to date there have not been many studies of the effectiveness of laboratory-based mathematics courses. The current research is intended to help provide information for those persons deciding whether to adopt a laboratory approach to calculus.

According to a recent survey (Tucker and Leitzel [18]) there is accumulating evidence, though still "spotty," that the broad goals of calculus reform are taking hold in the Nation's colleges and universities.

> ". . . the real significance of the calculus reform effort is the change that is occurring in collegiate classrooms across the country. Faculty and students are enthusiastic about the learning environments that the new approaches are creating." [18, p. vii.]

But there are detractors to calculus reform, to be sure. For example, Andrews [1, p.17] has stated,

> ". . . the technology aspect of calculus reform is especially disturbing. Many of our students have pitiful skills in arithmetic, algebra and trigonometry. While our brightest might gain much from *Mathematica* projects, we must be vigilant lest the B– or C+ students replace basic math skills with button pushing."

---

[1]For more details on the calculus reform movement, see *UME Trends*, Vol. 6, No. 6, January 1995, especially Ronald G. Douglas, "The first decade of calculus reform," pages 1 ff. and Alan H. Schoenfeld, "A brief biography of calculus reform," pages 3 ff.

As will be seen below, the status of student computational proficiency in the "new calculus" is addressed in the evaluation of various of the projects, including C&M.

Calculus Reform Projects (Selected)

Use of programming languages (Purdue University).

The Purdue University Project is more a reform of calculus pedagogy than of content. The centerpiece of the Project's philosophy is that students should construct their own understanding of mathematical concepts. Hence, the primary role of teaching is not to lecture, explain, or attempt to transfer mathematical knowledge, but to create situations for students that will foster their mental development. Under these assumptions new curricula were developed that combine a theoretical development with concrete applications and emphasize ideas as opposed to techniques of calculation. The mathematical programming language, *ISETL* (instructional set language), was adopted as the main tool for the Project because it involves students in a level of programming that is essential to the theory of learning. Interfaces have been created that make it possible for students to utilize *Maple* software in conjunction with *ISETL*.

During the first two years of the Purdue Project a comparative study was carried out in which students took common departmental final exams. The results indicated that, in comparison with the students in the conventional course, de-emphasizing mechanics did not put the Project students at a disadvantage. Furthermore, the rewards attending deeper understanding of calculus concepts seemed to outweigh difficulties that the Project students encountered (Schwingendorf and Dubinsky [13]).

Use of computer algebra systems (Duke University).

Project CALC (Calculus As a Laboratory Course) at Duke University utilizes an interactive computer laboratory approach that emphasizes writing and student cooperation as integral parts of class activity. Key features of Project CALC include: real-world problems, hands-on activities, discovery learning, the writing and revising of reports, group work, and using available tools. CALC classes meet for three 50-minute periods in a classroom equipped with one computer for instructor demonstration. Lecturing is limited to brief introductions to new topics and responses to requests for more information. Teams of four students work on substantial problems that lead to written reports approximately every other week. Each class splits into two groups; with each group scheduled for a weekly two-hour laboratory. Each lab team, consisting of two students, submits a written report almost every week. The principal software packages used are: *Derive* (for symbolic and graphic computation), *MathCAD* student edition (for numerical and graphical computation and for discovery experiments), and *EXP* (for technical word processing).[2]

---

[2]Now, this software has been replaced by using HP-48G calculators as the platform for all

In comparison with those traditionally taught, CALC students were better able to formulate mathematical interpretations of verbal problems and solve and interpret the results of some verbal problems. Classroom observers found that CALC students were much more actively engaged than were those traditionally taught. CALC students did less well than the traditional students on computational skills involving symbolic manipulation.[3] In follow-up surveys one and two years after the course, CALC students, significantly more than traditional students, reported that they better understood how mathematics is used and that the emphasis was on understanding mathematics rather than memorizing formulas (Bookman and Friedman [2]).

Calculus and *Mathematica*.

Of all of the calculus reform projects, Calculus and *Mathematica* (C&M) is probably the most technology-intensive. The course is based on an entirely interactive text in which the student has access to as many examples as desired. Through the use of technology, students see calculus as a course in scientific measurement, calculation, and modeling. Technology also makes it possible to present the subject as a highly visual, often experimental, scientific endeavor. (Uhl [20, p. 5])

The C&M course consists of four main sections: Introduction; Differential Calculus; Integral Calculus; Series and Approximation. C&M students are expected to attend 3 one-hour laboratory sessions every week. But in order to complete the work, they typically are in the laboratory between 6 and 12 hours per week.

Each C&M section has an instructor who has complete responsibility for that section. The instructor makes assignments and prepares examinations as s/he sees fit. The instructor also conducts a weekly discussion (not lecture) hour and spends other class hours visiting with students in the lab, doing individual teaching and responding to questions as needed. Instructor-student interaction is personalized and warm. Instructors are assisted in the lab by undergraduates recruited from students who previously took the C&M course. These lab assistants are available to help with technical aspects of the software and the mathematical content.

In the C&M laboratory, 30 computers are arranged as shown in Figure 1, with screens facing the center of the room.

Mathematica notebooks.

Students learn the course content through lessons called "notebooks" installed on the computer. The "notebook" is a tool of *Mathematica*, a powerful and easy to use computer algebra system. Each notebook opens with "basics" problems that introduce the key concepts, followed by "tutorial" problems in techniques

---

laboratory assignments.

[3]The project CALC evaluation report states that the Project faculty were of the opinion that computational skills could be improved by including more practice in routine calculations. (Bookman and Friedman [2])

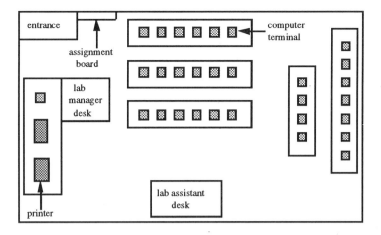

FIGURE 1. Arrangement of the C&M Calculus Laboratory.

and applications. Full solutions are available for both the basic and tutorial problems. The notebook closes with a "give it a try" section that contains problems to be solved. Students solve the problems using standard word processor and calculating software with graphic capabilities. These solutions become a key component of each student's notebook, which is electronically submitted to the instructor for comments and grading (Brown, Porta, and Uhl [3]).

Samples of Calculus and *Mathematica* course material.

Guided discovery learning.

In C&M, students are to acquire most of the rules and formulae by exploring using *Mathematica*. For example, the Chain Rule is introduced as follows:

> Let's check out the derivative of the composition of two functions: Here is the derivative of $\sin[x^2]$:
>
> *In [1]:* = **D[Sin [x ˆ 2], x]**
>
> *Out [1]:* = 2x Cos $[x^2]$
>
> This is interesting because the derivative of $\sin[x]$ is $\cos[x]$ and the derivative of x^2 is $2x$. It seems that the derivative of $\sin[x^2]$ is manufactured from the derivative of $\sin[x]$ and the derivative of x^2. Here is the derivative of $(x^2 + \sin[x])^8$:
>
> *In [2]:* = **D[(x ˆ 2 + Sin [x] ˆ 8, x]**
>
> *Out [2]:* = 8(2x + Cos[x]) $(x^2 + \text{Sin } [x])^7$
>
> This is interesting because the derivative of x^8 is $8x^7$ and the derivative of $x^2 + \sin[x]$ is $2x + \cos[x]$. It seems that the derivative of $(x^2 + \sin[x])^8$ is manufactured from the derivative of x^8, the derivative of $\sin[x]$ and the derivative of x^2.

The Chain Rule is taught using a kind of guided discovery. The students are not simply told the rule, $D[f[g[x]]] = f'[g[x]]g'[x]$. Instead, the C&M lesson guides the students through the examples leading up to a conclusion and lets them find the rules on their own. When students are not able to determine the rules with given examples, they try more examples and test ideas until they discover the rules.

Bottom-up (inductive) approach.

The cognitive learning procedure of the C&M group could be described as a bottom-up process. Learning the relationship between the shape (for example, 'peaks and valleys') of the graph of a function and the sign of its derivative is an example of the bottom-up process.

a. The $f'[x]$ graph represents the instantaneous growth rate. When $f'[x]$ is negative, the $f[x]$ graph is going down. When $f'[x]$ is positive, the $f[x]$ graph is going up.

b.

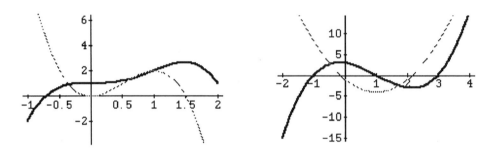

The solid line represents $f(x)$ and the dotted line represents $f'(x)$.

If students learn the rules (Part a above) first and then study the graphs (Part b above) in terms of the rules, this is a *deductive* or *top-down process*, in the sense that a knowledge representation structure has been retrieved from memory and is now guiding the process in solving a problem. On the other hand, making a decision on (a) from (b) is an *inductive* or *bottom-up process* because students learn mathematical principles by exploring examples and discovering the rules for themselves.

The C&M approach resembles an inductive process. Instead of providing the relationship between the shape of a function and the corresponding sign of its derivative, the students develop an intuition for the subject and work with examples until they get a conclusive idea. Accordingly, the direction of the cognitive process is from specific to general. Experimenting with a variety of examples on the same topic seems to induce certain analytical thought patterns in the students and to help them come up with generalizations on their own.

Method

The Department of Mathematics at the University of Illinois at Urbana-Champaign offers two versions of freshman calculus courses,[4] Calculus and *Mathematica* (C&M) and a standard course. C&M is a computer laboratory course with a weekly discussion session and no lectures. The standard course is lecture-oriented, making no use of computers. Students in both courses meet the same number of hours per week, but as noted earlier, C&M students allocate their time between classroom and laboratory.

Subjects.

The present study involved students enrolled in the second semester freshman calculus course at the University of Illinois at Urbana-Champaign during the spring of 1992. The experimental group consisted of two sections of C&M with a total enrollment of 26 students. The control group was made up of 42 students in a standard course. Enrollment in either the experimental or control group was by student self-selection.

Instruments.

Achievement test.

Achievement tests consisting of 16 items each were given at the beginning (pre-test) and at the end (post-test) of the semester. The pre-test was designed to measure knowledge of prerequisite content for introductory calculus while the post-test was intended to assess understanding of key concepts in the course. The content of the post-test was: differentiation (5 items), integration (6 items), and series and approximation (6 items). Both the pre- and post-achievement tests had 8 items at the conceptual understanding level and 8 items at the computational proficiency level. All items were in an open response ("show-all-work") format. To increase reliability of experimental results, a dual grading system was adopted: the grading was done and inspected by both the investigator and a teaching assistant. Before grading the tests, the investigator consulted with the teaching assistant to establish general guidelines as to how to distribute the points. To help ensure that the same criteria for grading were used, a detailed list was made that specified points for setting up the problem and for partial credit to be assigned for error types. Furthermore, both graders rated problem

---

[4]While this was true at the time that the research was carried out, the Department currently offers a third option, the Harvard Consortium Calculus, as well. Since the time of this study C&M has grown to include more than thirty sections per year.

1 for all students before going on to problem 2, and continued in that manner for the rest of the test. In most cases, the two scores for each student were the same, but if not, a compromise was found.

Attitude survey.

Pre- and post-attitude surveys consisting of the same items, with a 5-point Likert scale (strongly agree to strongly disagree), were used. These items, adapted from the questionnaire used in the Second International Mathematics Study [17], were based on (1) attitudes toward mathematics, which had four dimensions: (a) mathematics as a process, (b) mathematics and affect, (c) cooperative learning, and (d) value to society; and (2) attitudes toward computers.

Concept maps.

In order to assess conceptual understanding, concept maps were used. Concept maps are two-dimensional graphic representations of concepts, propositions, and their relationships. They are graphic organizers that represent content diversity, super ordinate-subordinate relationships, and interrelationships among subordinate concepts [12]. Constructing concept maps requires students to externalize their thinking by mapping out their conceptual structure of a topic. Figures 2a and 2b are simple examples of concept maps.

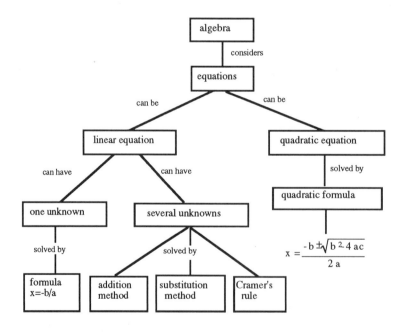

FIGURE 2A. A concept map for algebra

At the end of the semester, concept map sheets (See Appendix I) with two examples (Figures 2a and 2b) were given to the students. The investigator then

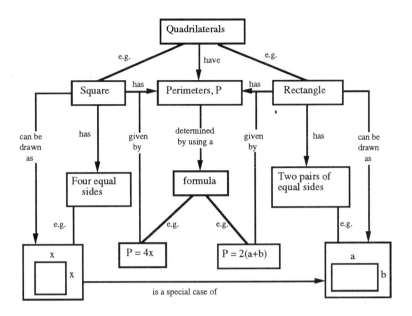

FIGURE 2B. A concept map for quadrilaterals

briefly explained how to construct a concept map and asked the students to
generate their own.

Student interviews.

The investigator interviewed the students in the one section ($N = 12$) of the
C&M group and one section ($N = 22$) of the standard group concerning their
understanding of key concepts in calculus.

Data analysis.

Achievement and attitudes.

The achievement data were analyzed using analysis of covariance (ANCOVA).
ANCOVA, which combines analysis of variance with regression analysis, is particularly useful when subjects (in this case, students) are not randomly assigned
to groups. In the present experiment, the C&M group and the standard group
were intact, not randomized, groups. Thus, the students might have inherent
differences in attitude and achievement that could lead to bias in the subsequent
analyses.

ANCOVA requires obtaining data on the dependent variable and its covariates. The covariates represent a pre-existing variation that has not been controlled in the experiment and is believed to affect the dependent variable. In
the present study, the pre-test on prior knowledge of mathematics (described in
'Instruments'section above) was used as the covariate.

Concept map analysis.

The student concept maps were analyzed by two methods that were devised by the first author. The maps were first scored on: propositions (2 points for meaningful propositions); hierarchy and cross links (5 points each for valid hierarchy and for significant cross links); concepts (3 points for meaningful concepts beyond those provided); misconceptions (deduct 1, 3, or 5 points according to extent of misconception). (See Figure 3.)

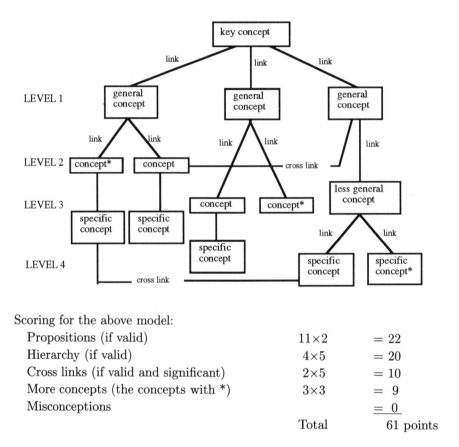

Scoring for the above model:

Propositions (if valid)	11×2	= 22
Hierarchy (if valid)	4×5	= 20
Cross links (if valid and significant)	2×5	= 10
More concepts (the concepts with *)	3×3	= 9
Misconceptions		= 0
	Total	61 points

FIGURE 3.   Scoring rubric for a concept map

The second mode of concept map analysis involves examining the congruence coefficients[5] between the concept map of the instructor and his/her students' concept maps. The range of congruence coefficients is 0 to 1, with the higher congruence coefficient indicating that the student concept maps have more similarity to the instructor's concept map. In this analysis, one matrix is constructed

[5]The software *MicroQAP*, a program for the computation of the generalized measure of association between the two data matrices, was used (Costanzo and Lawrence [4]; Hubert et al. [7] and [8]).

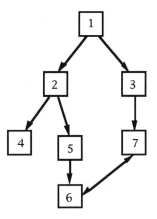

FIGURE 4.   Simple example of a concept map

from the instructor's concept map and the other is constructed from a student's concept map. Figures 4 and 5 show how the matrices are created.

In the concept map in Figure 4, $1 \rightarrow 2$, $2 \rightarrow 4$, $2 \rightarrow 5$, $5 \rightarrow 6$, $1 \rightarrow 3$, $3 \rightarrow 7$, $7 \rightarrow 6$, and $6 \rightarrow 7$. The corresponding matrix is shown in Figure 5.

	1	2	3	4	5	6	7
1	0	1	1	0	0	0	0
2	0	0	0	1	1	0	0
3	0	0	0	0	0	0	1
4	0	0	0	0	0	0	0
5	0	0	0	0	0	1	0
6	0	0	0	0	0	0	1
7	0	0	0	0	0	1	0

FIGURE 5.   Matrix created from the concept map in Figure 4

Results

Achievement.

When the results of the end of semester calculus tests were adjusted (using ANCOVA) for students' prior mathematical knowledge, the C&M scores were found to exceed ($p < .05$) those of the standard group, suggesting that Calculus and *Mathematica* had a positive influence on student achievement (see Table 1). Similarly, when the scores were analyzed according the level of conceptual understanding, those for the C&M Group were greater ($p < .05$) than for the standard group. No statistically significant difference between the two groups was found for computational proficiency (see Table 2).

The results of the achievement test suggest that the C&M group attained a higher level of conceptual understanding than the standard group without loss

TABLE 1. Summary Table of ANCOVA
for the post-achievement test (total)

Source	SS	Df	MS	F	p
Covariate	3222.3	1	3222.3	119.8	.001
Treatment	120.7	1	120.7	4.5	.038
Within	1748.6	65	26.9		

TABLE 2. Comparison of the C&M group and a standard
group on the post-achievement test; by cognitive level of items

Group (N)	Conceptual Understanding		Computational Proficiency		Total	
	Mean	SD	Mean	SD	Mean	SD
C&M (26)	29.5	4.7	28.9	6.0	58.4	8.9
Standard (42)	24.9	5.3	29.8	5.6	54.7	8.5
	$F = 16.4$		$F = 0.9$		$F = 4.5$	
	$p = .0001$		$p = .3392$		$p = .0380$	

of computational proficiency. Of special interest was the finding that the items contributing to the greatest difference between the C&M and standard group were those concept-oriented items related to graphs and those having to do with the meaning of theorems (see items in Appendix III).

Attitude.

When the attitude scores at the end of the semester were adjusted for prior differences, those of the C&M group were found to be more positive ($p < .05$) than the standard group. (See Table 3.) In particular, there were substantial differences in attitudes toward computers, cooperative learning, and mathematics as a process. (See items in Appendix II.)

TABLE 3. End of semester attitude
data for the C&M and standard groups

Group (N)	Attitudes Toward Mathematics		Attitudes Toward Computers		Total	
	Mean	SD	Mean	SD	Mean	SD
C&M (26)	71.7	4.6	20.2	1.9	91.8	4.9
Standard (42)	68.4	6.0	17.4	2.4	85.8	6.3
	$F = 9.2$		$F = 0.5$		$F = 19.2$	
	$p = .0035$		$p = .0001$		$p = .0001$	

Results of Concept Map Analysis

Cross links.

An analysis of the concept maps revealed that the scores of the C&M group were generally higher than those of the standard group. Especially distinctive differences were found for the cross links. For example, more students in the C&M group than the standard group made the following connections: derivative and integral; differentiation and integration by the Fundamental Theorem of Calculus; power series and geometric series. Cross-link scores also favored the C&M students for interpretation of integration-by-parts as the Product Rule of differentiation, and for explanation of integration-by-substitution as the Chain Rule of differentiation. (See Table 4.)

TABLE 4. Concept map scores for C&M and standard groups

| Criterion | Concept map (A) | | | | Concept map (B) | | | |
| | C&M | | Standard | | C&M | | Standard | |
	Mean	SD	Mean	SD	Mean	SD	Mean	SD
Propositions	29.8	0.7	27.2	2.8	15.6	0.9	15.2	1.0
Hierarchy	24.4	3.9	27.7	5.9	25.0	3.5	23.0	5.9
Cross links	10.0	5.0	5.0	3.8	3.3	2.5	1.3	2.3
More Concepts	7.0	4.7	10.8	10.6	9.3	4.4	4.0	4.6
Misconceptions	-1.4	2.2	-2.3	3.3	-1.3	3.0	-2.8	2.3
Total	69.8	11.1	68.4	13.5	51.9	9.1	40.7	7.4

Congruence between instructor and student.

The C&M student concept maps showed greater congruence with their instructor's concept map than did the standard group with their instructors. See Table 5.

TABLE 5. Congruence between
instructors' and students' concept maps

| Group | Concept Map (A) | | Concept Map (B) | |
	Mean	SD	Mean	SD
C&M	0.66	0.14	0.58	0.10
Standard	0.53	0.11	0.37	0.18

A strong positive correlation ($r = .73$, $p < .01$) was found between cross links and total achievement test score, suggesting that students who did well in linking concepts in different branches generally received high achievement test scores. Furthermore, most of the students in both groups tried to include more concepts beyond the given lists. The two groups differed substantially, however, in the

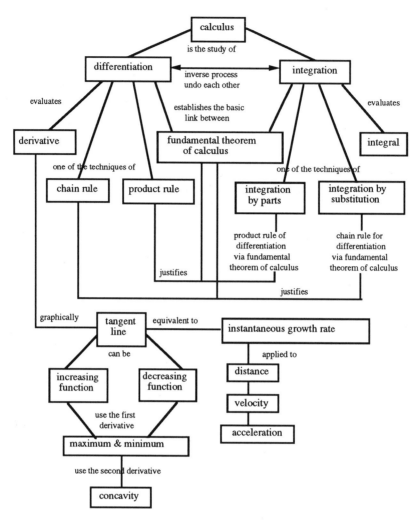

Scoring: Propositions 15×2 = 30
          Hierarchy 6×5 = 30   Congruence
          Cross links 3×5 = 15   Coefficient:
          More concepts 4×3 = 12   .88
          Misconceptions   = 0
                             Total   87

FIGURE 6. High scoring map demonstrating the understanding of interrelationship between the concepts

concepts chosen by the two groups. Most of the concepts selected by the C&M group were related to graphs, visual interpretation and examples.

By contrast, most of the concepts chosen by the standard group were terms, applications, and techniques. Figures 6, 7 and 8, 9 indicate the differences in the complexity of the high and low scoring maps. The two concept maps constructed by the C&M group (Figures 6, 7), even though not perfect in the concept listing and connecting lines, demonstrate a tendency to use more concepts and to list relevant interrelationships between concepts. The two concept maps produced by a standard group (Figures 8, 9), on the other hand, show a relatively simplistic view of the calculus concepts without reference to cross links.

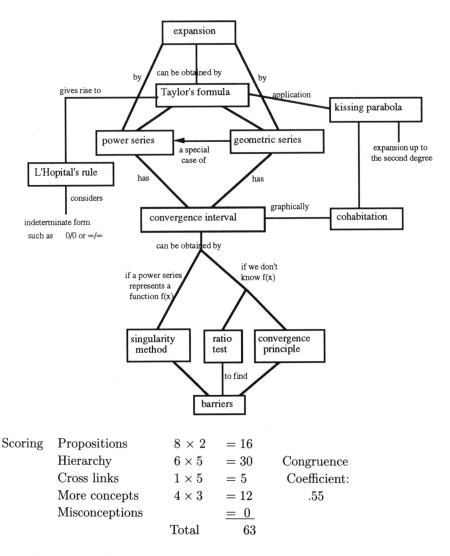

	Scoring			
Scoring	Propositions	8×2	$= 16$	
	Hierarchy	6×5	$= 30$	Congruence
	Cross links	1×5	$= 5$	Coefficient:
	More concepts	4×3	$= 12$	.55
	Misconceptions		$= 0$	
	Total		63	

FIGURE 7.   High scoring map demonstrating large number of identified relevant concepts

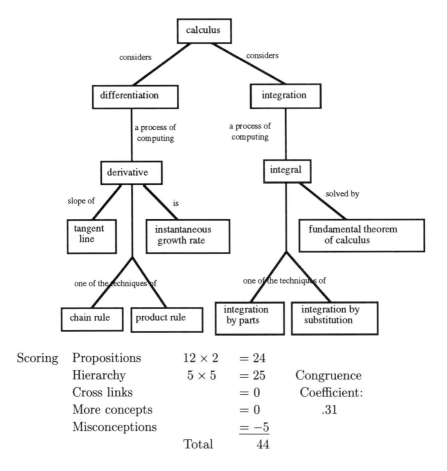

Scoring   Propositions 12×2 $= 24$

         Hierarchy 5×5 $= 25$ Congruence

         Cross links $= 0$ Coefficient:

         More concepts $= 0$ .31

         Misconceptions $= -5$

                         Total   44

FIGURE 8.   Low scoring map showing lack of interrelationships

Student interviews.

Among the six questions asked, responses to the following two questions showed distinctive differences between the C&M and standard group.

Question 1.

"Does the slope of the tangent line to $f(x)$ at point $(a, f(a))$ represent the derivative of the function f at the same point?"

Most of the C&M group (10 out of 12 students) answered Question 1 affirmatively, while the standard group was about evenly split (11 and 11 students). The different responses might be explained in the following way: the standard group was first introduced to the derivative via the slope of the tangent line, and then taught the more formal epsilon-delta (limit) definition of the derivative. They identified the derivative as the slope of the tangent line rather than by the complicated epsilon-delta definition. Typical examples of the tangent line in the textbooks resemble Figure 11. Thus, the standard group likely was not familiar with the graph in Figure 10 as a tangent line. These factors probably caused the

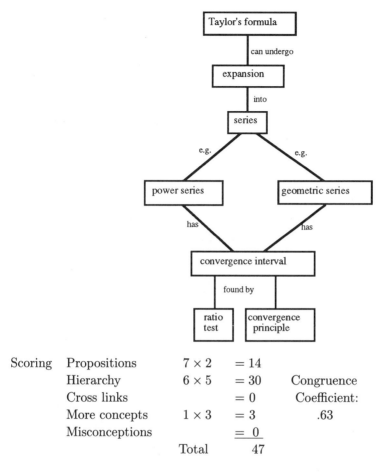

Scoring   Propositions 7×2 $= 14$
           Hierarchy 6×5 $= 30$ Congruence
           Cross links $= 0$ Coefficient:
           More concepts 1×3 $= 3$ .63
           Misconceptions $= 0$
                              Total   47

FIGURE 9. Low scoring concept map demonstrating relatively simplistic view with few relevant concepts

students to reply that the slope of the line did not represent the derivative in the graph in Figure 10. The C&M students, on the other hand, were taught neither the derivative as the slope of the tangent line or the epsilon-delta definition. Instead, they recognized the derivative as an instantaneous growth rate.

Question 2.

"Is the integral of a function over a closed interval a number or a function?"

Most of the C&M group (11 out of 12 students) answered this correctly. By contrast, fewer students in the standard group (15 out of 22 students) gave the right answer. The probable reason is that the real difficulty in hand calculation of a definite integral lies in finding an antiderivative. Thus a standard group is likely to lose the central idea, and answer that the integral of a function over a closed interval is a function. However, the C&M group typically can get the result of integration quickly and easily by *Mathematica* commands. Therefore, they are apt to consider

$$\int_a^b f(x)\,dx$$

as a number.

Generally, the C&M group seemed to more clearly understand the nature of the derivative and the integral than did the standard group. The simple definition of the derivative as an instantaneous growth rate and the direct introduction to the definite integral might have promoted their understanding.

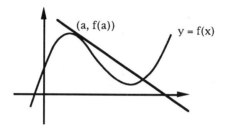

FIGURE 10. Graph used in interview question 1

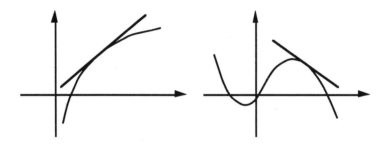

FIGURE 11. Examples of the tangent line

Conclusions

The advantages of the C&M course over the standard course include: visualization of ideas, discovery learning by exploration, and cooperative learning. Visualization exploits the power of *Mathematica* to help promote sound conceptual understanding. Most of the topics in the C&M courseware are coordinated with *Mathematica* graphics. Thus, C&M students are helped to visualize the intended mathematical concepts. For example, visualization by *Mathematica* seems to be an effective tool for learning the derivative by the difference quotient, the Trapezoidal Rule, and the convergence interval. It also appears that the experimentation promoted by *Mathematica* leads to redoing, reformulating, rethinking, adapting, and making changes. Students are therefore assisted in discovering such basic concepts and principles as the Chain Rule, the Fundamental Theorem of Calculus, and the relationship between the shape ("peaks and valleys") of a function and the sign of its derivative. A desirable side effect of C&M was the rapport that was established among the students in the laboratory setting. Considerable evidence was obtained that when students gathered around the computer, worked together, and shared and developed ideas, a great deal of mathematics was learned.

One of the most frequently discussed disadvantages of computer-based approaches to mathematics is that of the black-box syndrome—that is, students blindly using computer courseware without understanding the underlying concepts and procedures (Kenelly and Eslinger [9]). The black-box syndrome might be prevented by carefully sequencing the examples and exercises of the courseware. For example, if a certain problem requires a somewhat different command line from that of the previous problem, then students can not merely copy the previous line. They should instead appropriately revise the command line. The courseware might also include questions as to why specific commands are chosen and how they work—thus causing students think about the meaning of numerical and graphic results.

It is believed by some that a laboratory course in calculus is too time consuming, and that students can become overly dependent on the software. But the current research has found that the C&M course allows the students to spend less time on computation and better direct their efforts to conceptual understanding. Accordingly, there was an increase in the students' conceptual attainment without a statistically significant loss in computational achievement. Moreover, the C&M approach appeared to produce improvement in the students' attitudes toward mathematics by relieving them of the tedium and source of errors involved with computations. There was also an improvement in attitudes toward the computer by making the students more comfortable with the machine and became acquainted with some of its capabilities. The calculating and plotting capabilities of *Mathematica* helped the students discover and test mathematical results in much the same way that a physics or chemistry student uses the laboratory to discover and test scientific laws. Those capabilities provided opportunities for students to consider more open-ended questions and to encounter more realistic problems than those often found in traditional calculus texts. In a word, the *Mathematica* courseware seemed to provide a challenging course in which the students learned a more lively calculus.

Limitations

There were two major limitations to this study. First, each experimental group was an intact group. That is, the samples were simply available classes, not random ones. A review of the classes in a search for possible bias revealed no difference between the two groups in the percentage of each gender, mean age and number of mathematics courses taken at the college level. But there was substantial difference in the distribution of majors between the two groups: 73% and 4% of the C&M group and 57% and 19% of the standard group majored in engineering and business administration, respectively. However, the investigators believe that this difference was not a major factor in the results.

As noted earlier, self-selection, on the part of both students and instructors, was also at work. Before enrolling in the C&M course, the students knew that the teaching method of the course was different from the usual courses. Thus, the members of the C&M group might have been more energetic and adventurous than those in the standard group. In the same way, the instructors who volunteered for the C&M course might have been more committed. The instructors worked hard. The students noticed and appreciated this commitment and as a consequence probably took the course more seriously.

The limitations and concerns all need to be considered before generalizing about the results of this study to other populations.

REFERENCES

1. Andrews, G. E., *The irrelevance of calculus reform: Ruminations of a sage-on-the-stage*, UME Trends **6(6)** (1995), 17, 23.
2. Bookman, J. and Friedman, C. P., *Final Report: Evaluation of Project CALC*, (mimeo) (1994).
3. Brown, D., Porta, H., and Uhl, J., *Calculus & Mathematica, courseware for the nineties*, *Mathematica* Journal **1(1)** (1990), The Mathematical Association of America, Washington, DC, 43–50.
4. Costanzo, C. M. and Lawrence, J. H., *A higher moment for spatial statistics*, Geographical Analysis **15(4)** (1983), 347–351.
5. Douglas, R. G., *Toward a Lean and Lively Calculus*, Mathematical Association of America, Washington, DC, 1986.
6. Heid, M. K., *An exploratory study to examine the effects of resequencing skills and concepts in an applied calculus curriculum through the use of the microcomputer*, (Doctoral dissertation, University of Maryland) Dissertation Abstracts International **46, 1548A** (1984).
7. Hubert, L. J., Golledge, R. G., and Costanzo, C. M., *Generalized procedures for evaluating spatial auto correlation*, Geographical Analysis **13(3)** (1981), 224–233.
8. Hubert, L. J., Golledge, R. G., Costanzo, C. M., and Gale, N., *Tests of randomness: unidimensional and multidimensional*, Environment and Planning **17(3)** (1985), 373–385.
9. Kenelly, J. W., and Eslinger, R. C., *Computer algebra systems*, Calculus for a new century (L. A. Steen, ed.), Mathematical Association of America, Washington, DC, 1988, pp. 78–79.
10. Leinbach, C., *The Laboratory Approach to Teaching Calculus*, Mathematical Association of America, Washington, DC, 1991.
11. National Science Foundation Program Announcement, *Calculus*, Washington, DC, 1987.
12. Novak J. D., and Gowin, D. B., *Learning How to Learn*, Cambridge University Press, Cambridge, 1985.
13. Schwingendorf, K. E. and Dubinsky, E., *Purdue University*, Priming the Calculus Pump: Innovations and Resources (T. W. Tucker, ed.), Mathematical Association of America, Washington, DC, 1990, pp. 175–198.
14. Small, D. and College, C., *Calculus reform—laboratories—CAS's*, The Laboratory Approach to Teaching Calculus (C. Leinbach, ed.), 1991, pp. 9–13.
15. Smith. D. A. and Moore L. C., *Duke University*, Priming the Calculus Pump: Innovations and Resources (T. W. Tucker, ed.), Mathematical Association of America, Washington, DC, 1990, pp. 51-74.
16. Steen, L. A. (Editor), *Calculus for a New Century: A Pump, Not a Filter*, Mathematical Association of America, Washington, DC, 1987.

17. Travers, K. J. (Editor), *Second International Mathematics Study: Detailed Report for the United States*, Stipes Publishing Company, Champaign, IL, 1986.

18. Tucker, Alan C. and Leitzel, J. R. C. (Eds.), *Assessing Calculus Reform Efforts: A report to the community, MAA Report 6*, Mathematical Association of America, Washington, DC, 1995.

19. Tucker, T. W., *Priming the Calculus Pump: Innovations and Resources* (1990), Mathematical Association of America, Washington, DC.

20. Uhl, J. J., *Calculus and Mathematica*, UME Trends **6(6)** (1995), 15.

Appendix I: Concept map sheet

Construct your own concept map with the concept list below.

Tips: Place the most inclusive concept at the top and show successively less inclusive concepts at lower positions on a hierarchy. Then, specify the appropriate linking words which indicate the relationships between concepts. Concept maps are idiosyncratic. Your concept map may not be similar to the concept map proposed by others; but they both may be correct and valuable. Including more concepts which are not given, and making cross links between the concepts in other branches deserves extra credit.

Concept List (A)

Calculus	
Differentiation	Integration
Derivative	Integral
Instantaneous growth rate	Tangent line
Chain rule	Product rule
Integration by parts	Integration by substitution
Distance	Velocity
Acceleration	Fundamental theorem of calculus

Concept List (B)

Expansion	Power series	Geometric series
Convergence interval	Ratio test	Convergence principle
Taylor's formula	L'Hopital's rule	

Appendix II: Attitude survey items favoring the C&M group

Attitude toward computers.

Using a computer makes learning mathematics more mechanical and boring.*
Using a computer can help you learn many different mathematical topics.
If you use a computer, you don't have to learn to compute.*
Solving word problems is more fun if you use a computer.

Cooperative learning.

I like to solve problems by working with others.
When I do mathematics with other students, I realize I am not the only one who can't understand.

Mathematics as a process.
Most of the learning of mathematics involves memorizing.*
Most mathematics problems can be solved in different ways.
*Because these items were negatively worded, scoring was flipped.

Appendix III: Achievement test items favoring the C&M group

1. What is the integration by parts formula and how is it related to the product rule of differentiation (i.e., $(fg)' = f'g + fg'$)?
2. Six functions are plotted below. Three of them are derivatives of the other three. Match the plot of each derivative with its corresponding function.

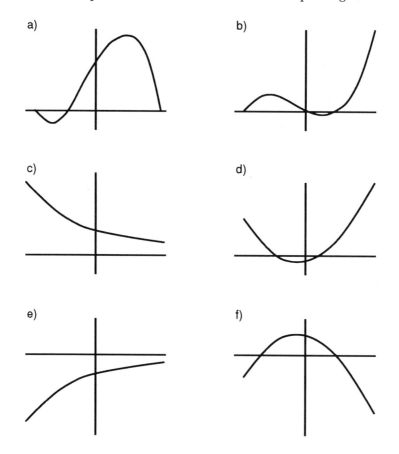

KOREAN EDUCATIONAL DEVELOPMENT INSTITUTE

UNIVERSITY OF ILLINOIS AT URBANA-CHAMPAIGN

CBMS Issues in Mathematics Education
Volume **6**, 1996

Differential Patterns of Guessing and Omitting in Mathematics Placement Testing

ALVIN BARANCHIK AND BARRY CHERKAS

ABSTRACT. In a study of over 1000 students at an urban four-year college, where major ethnic and language minority groups are broadly represented, the traditional method of scoring—by totaling the number of correct answers—on a standardized multiple-choice placement test for a course in precalculus was investigated. It was common to find differential patterns of guessing and omitting by ethnicity, first language, age, gender and birthplace. While mathematics skill seemed to account for variation across gender and age groups, it did not explain differential patterns across ethnic, first language and birthplace groups. Further, when studying differential patterns of not finishing the test, gender variation was observed in certain ethnic and age groups and this could not be attributed to differences in mathematics skill. Thus, number correct scoring appeared to lessen the chance that members of certain cultural and gender groups gain admission to a gateway course for science-related majors, for reasons not entirely mathematical in nature. Formula scoring, with its correction for guessing and omitting, was found to counter this negative effect of number correct scoring by increasing the ranking of a typical omitter's score by over three percentile points. To enhance fairness in mathematics placement testing, aside from remedies that utilize formula scoring, several steps to strengthen test directions and procedures are proposed consonant with traditional number correct scoring.

Introduction

Placement testing in mathematics is an essential part of advising, screening and managing entry level students at colleges and universities with heterogeneous populations. It gained status within the collegiate mathematics community when, in 1977, the Mathematical Association of America developed a battery of placement tests and established the Committee on Placement Testing (now, the Committee on Testing) to administer the Placement Testing Program [26].

Of the 3600 colleges and universities in the United States, over 460 subscribe to the Mathematical Association of America Placement Testing Program [27]; many others use either commercially prepared or locally developed tests.

At colleges that provide remedial or developmental education, testing is often used to mandate placement into either (a) college level mathematics with full college credit or (b) pre-college level mathematics with little or no college credit. Due to the involuntary nature of such placement, questions of fairness take on added importance, especially at urban institutions where culturally diverse populations are common. The Mathematical Sciences Education Board has stated the Equity Principle: "Assessment should support every student's opportunity to learn important mathematics" [29]. In addition, the Board has called for further research into assessment issues. In this paper, we examine non-academic factors that may work against fairness in mathematics placement testing.

The placement test is usually in a convenient-to-grade multiple-choice question format that is conventionally scored by totaling the number of correct answers (number correct scoring). Inasmuch as students are at liberty to choose different strategies, Mislevy and Verhelst [31] have called into question the validity of relying upon number correct scoring. Evidence that supports this view for placement testing comes from Ben-Shakhar and Sinai [6] who reported that females showed a greater tendency to omit answers and a lesser tendency to guess than did males, despite instructions to all students that guessing was not penalized. To measure guessing and omitting tendencies, they used special indices [5, 42], which we describe in the section titled "Measures of Guessing, Omitting and Not Finishing." Daunis [14] and Hudson [17] reported finding gender bias in mathematics placement scores, each noting that in subsequent course work, females outperformed males with comparable placement scores; that is, placement scores underpredicted female course performance. This phenomenon led Daunis to suggest consideration of an unusual remedy: different cutoff scores for men and women.

An important objective of mathematics placement testing is to distinguish between adequate and inadequate levels of students' mastery of prerequisite mathematical knowledge. When varying patterns of guessing and omitting are left unaccounted-for, as in conventional number correct scoring, similar scores may mask different levels of mathematical skill. Users who rely on number correct scoring may harbor a belief that omitted answers (or a reluctance to guess) are simply a manifestation of—and an indication of—weak mathematical skills. But if differences in guessing and omitting are associated with non-academic factors such as ethnicity, language or gender, the objectivity of number correct scoring is open to challenge. Some might minimize this controversy by noting that the test score is only supposed to be one factor in making placement decisions; indeed, it is sound practice to consider information from several sources, for example, high school record [28]. However, at institutions with scarce resources, the test score can be decisive.

In this paper, we examine a culturally diverse population and study differential patterns of guessing and omitting on a college mathematics placement test for various demographic variables: ethnicity, first language, age, gender and birthplace. We address the following research questions:

(1) To what extent do patterns of guessing and omitting vary across such demographic variables as ethnicity, first language, age, gender and birthplace?

(2) Can variation in one demographic variable be explained by variation in the other demographic variables, or be accounted for by differences in mathematics skill?

(3) Is the tendency to not finish the test present in any particular demographic group?

(4) What are the implications of using formula scoring (which incorporates a correction for guessing or for omitting) in place of conventional scoring?

Research into mathematics placement testing is common at the dissertation level. Quantitative studies appear regularly; most studies attempt to determine predictive validity. Many seek to identify correlates and predictors of success in entry-level college mathematics, such as gender, ethnicity, age and mathematics anxiety (see viz.: [21, 30, 32, 39]). The general tenor of these dissertations is that subsequent performance of women and returning adults is often found to be underpredicted while that of blacks is found to be overpredicted.

Apart from dissertations, the research literature related to demographic differences in mathematics placement testing is limited. Gender differences were reported in scores on mathematics placement tests in a study of nine unrelated universities [8]. Differences in placement scores between blacks and whites were reported in a study at a midwestern commuter university [23]. In a six college study, Kanter [20] found wide disparities in the use of mathematics placement tests that may affect access to college courses for ethnic groups that are underrepresented in mathematics-related fields. We found a lack of research on language minority differences in mathematics placement testing.

Beyond placement testing, gender and minority differences were commonly reported in scores on the mathematics section of the Scholastic Assessment Test (SAT) [10, 34, 41]. Broader performance issues in mathematics by gender were meta-analyzed by Hyde, Fennema, and Lamon [18] and surveyed by Leder [22], who also pointed to "considerable evidence that females make more omissions errors than males." Substantial patterns of minority performance dissimilarities in K-12 mathematics were documented in the review by Secada [36]. A bibliography on assessment issues for language minorities was published by Fairtest [15].

Number Correct and Formula Scoring

Guessing and omitting on multiple-choice questions create a thorny scoring predicament [4, 40]. The issue is usually skirted on placement and in-class tests, partly because the standard corrections for guessing or omitting are commonly perceived as penalties or gifts, respectively, and partly because number correct scoring is easier to administer.

On a multiple-choice test with n questions and k choices per question, the formula for scoring with a correction for guessing is given by

$$S_G = c - \frac{w}{k-1},$$

where c is the number of correct answers and w is the number of incorrect answers. The derivation of this formula makes use of the crude assumption that all incorrect answers on a multiple-choice test are the result of random guessing [40]; here, $\frac{w}{k-1}$ estimates the number of correct answers resulting from random guessing, since, on average, the ratio of correct to wrong guesses is 1 to $k-1$. Alternatively, the formula for scoring with a correction for omitting is given by

$$S_0 = c + \frac{b}{k},$$

where b is the number of blanks $(n = c + w + b)$. This formula "assumes that an individual omitting an item should receive the expected weight under conditions of random response to that item" [33].

Despite the dissimilar assumptions underlying these two modes of formula scoring, the rankings they produce, and hence the percentile scores, are identical. To see why, use $w = n - c - b$ to rewrite the formula for S_G as

$$S_G = \frac{k}{k-1}c - \frac{n-b}{k-1} = \frac{k}{k-1}S_0 - \frac{n}{k-1}.$$

Thus, S_G is monotonically increasing in S_0 and the scoring orders are identical.

If there are no omissions at all, formula scoring using either S_G or S_0 is superfluous because it then produces a ranking identical to that of number correct scoring. Within the measurement community, debate over when and where to use which scoring method, "Number Correct versus Formula Scoring," has persisted over 60 years [9]. Currently, most large-scale testing programs administered by the Educational Testing Service (ETS) use the formula that corrects for guessing; this includes the SAT, the College Board Achievement Tests and the Graduate Record Examination (GRE) subject examinations. However, there are several other large-scale testing programs that use number correct scoring, including the American College Testing and the GRE general examinations. There is no large-scale testing program that uses the formula that corrects for omissions.

Test Directions and Procedures

. . . it is likely that there will always be some examinees who fail to listen to instructions, fail to believe them, or fail to understand any test-taking strategy. It may be impossible to construct any scoring system or set of instructions that is totally suitable to the entire examinee population. (Grandy [16])

The scoring of multiple-choice tests is inextricably linked to the strategic orientation of examinees. At the extremes, some test takers may omit all questions they do not know, while others may adopt the opposite strategy of guessing on all questions they do not know. Other strategies might include guessing only when at least one choice can be eliminated, or selecting a middle or longest answer.

The MAA Placement Test Program User's Guide [28] is silent on the issue of telling students how the test is scored, and the instructions printed on the MAA tests do not explain how the test is scored. This approach is consistent with keeping a focus on the subject matter being tested, yet also presumes that the resultant scores are free from errors.

The standard practice of ETS is to inform examinees about the scoring rules prior to the test in handout brochures. This is part of a larger effort to establish uniform test conditions for all examinees. At the test site, ETS scoring instructions are printed on the tests and repeated by the test administrator as well. Each scoring method has its own set of directions, whose descriptions follow.

Number correct scoring is accompanied with instructions to examinees stating that there is no penalty for incorrect answers; examinees are further directed to both watch their time and answer all questions. This latter advice is objectionable on educational grounds because, by discouraging omissions, it encourages indiscriminate guessing and focuses student attention on improving their score without regard to their knowledge of the subject. Such conduct lowers the reliability of test scores. There is, in fact, evidence that indiscriminate guessing occurs with these instructions, especially toward the end of a test, where "Numerous instances have been observed of 'pattern-marking' the last several items [. . .] indicating clearly that random responding has taken place" [5].

Formula scoring is accompanied by instructions to examinees stating that a fraction of the incorrect answers will be subtracted from the number of correct answers; examinees are further directed to watch their time, to guess only if they can eliminate one or more options, and in addition they are told that random guessing is not likely to change their score significantly. This advice is objectionable on educational grounds because it draws attention away from the subject matter being tested, and instead focuses examinees on question-by-question strategizing to avoid what is perceived as a penalty. There are also indications that the additional obligation to consider and weigh each response requires more time on the part of examinees [2, 3, 7]. Of additional concern is evidence that examinees who are reluctant to guess score lower than examinees with comparable partial knowledge who are not reluctant to guess [37, 38]. Further research has clarified that while this holds generally for higher scoring examinees, who may be expected to have partial knowledge, the opposite holds for lower scoring examinees due, perhaps, to greater amounts of misinformation [5].

Method

Institutional background and curriculum.

We gathered data for this study over four consecutive semesters—fall 1991, spring 1992, summer 1992 and fall 1992—at a four-year liberal arts "subway" college that is part of a comprehensive public university with an open admissions policy. Before the fall 1991 semester, the college offered a two-semester sequence of precalculus courses. Although undergraduates were not mandated to take mathematics, several departments required the first semester of precalculus for their majors. There were no college prerequisites for precalculus beyond passage of a university-wide skills assessment test and college proficiency examination, both in arithmetic and the beginning stages of algebra. Since the first precalculus course had a successful completion rate of only 45% (Success = {A, B, C or Credit}) and a withdrawal rate of 18%, the mathematics department decided to add to the curriculum an intermediate algebra course and to develop an algebra placement testing program. Due to limited resources, they selected a single test: the 35 question, 30 minute Elementary Algebra Skills Test [11].

The department piloted the test in fall 1991, unannounced, to all students who appeared on the first day of classes in the first precalculus course and the scores were later analyzed to set a cutoff score on the basis of subsequent course success. By fall 1992, the placement test was being given before classes began and students had an opportunity to prepare for it if they wished, but they needed to pass the test or successfully complete the intermediate algebra course to register for precalculus. The high rates of failure and withdrawal, which motivated this procedure, were somewhat mitigated: in the year following full implementation, the rate of satisfactory completion increased to 62% and withdrawals dropped to 9%.

Population.

There were 1003 students who took the placement test. To statistically control for prior mathematics skill, we used results from a standardized test in basic mathematics skill [13] given to all entering students (college proficiency test). Additional population data on age, ethnicity, gender and major were obtained from school records; data on birthplace and first language were obtained from a separate questionnaire filled out at the same time as the placement test. Of the 40% of students in this study who had declared a major or premajor, 29% selected science and mathematics-related fields; 23%, health sciences; 33%, nursing; and 15%, other fields. Other population characteristics—ethnicity, first language, age, gender, birthplace and prior mathematics skill—are summarized in Table 1.

TABLE 1. Population characteristics

Group	Subgroup	N	Percentage	Prior Math Skill N	Mean[a]
Ethnicity	Asian	208	22.1	205	36.0
($N = 943$)	Black[b]	265	28.1	259	33.2
	Latino	199	21.1	193	33.0
	White	271	28.7	242	37.5
Language	Chinese	89	11.9	87	36.6
($N = 748$)	English	490	65.5	457	36.4
	French	32	4.3	28	31.5
	Spanish	137	18.3	133	31.7
Age	≤ 20	429	42.8	428	36.6
($N = 1003$)	21-24	284	28.3	280	33.7
	≥ 25	290	28.9	245	33.6
Gender	Female	696	69.4	661	34.3
($N = 1003$)	Male	307	30.6	292	36.6
Birthplace	USA	472	50.1	437	36.8
($N = 943$)	Foreign	471	49.9	460	33.5

[a]Number correct out of 45   [b]Non-Latino

Instrument

Data on the measurement attributes of the mathematics placement test used [11], including statistical reliability, appears in the associated guide [12]. Each question included four choices. We scored the test and ranked students in two ways: number correct scoring and formula scoring.

Test directions and procedures.

In addition to a handout describing the test but not the scoring rules, at the time of the examination students were directed to read and invited to ask questions about the following instructions printed on the back cover of the test [11]:

> You should mark an answer to every question even if you are not sure of the correct answers for some of them. Your score will be the number of correct answers. No credit will be taken away for incorrect answers.

Test administrators did not repeat these rules, nor did they remind examinees towards the end of the test about how much time remained.

Measures of Guessing, Omitting, and Not Finishing

Differences in group patterns of guessing and omitting were studied using two measures of guessing, three measures of omitting and a measure of not finishing the test. Omitting can be measured directly from examinee responses.

To this end, we define the intermediary variables: (1) O_1, the total number of questions omitted (left blank) by an examinee and (2) O_2, the total number of questions omitted before the last question answered by an examinee. Because the highly skewed characters of $\{O_i\}$ make them less amenable to standard statistical analysis, we consider instead the corresponding 2-valued (dichotomous) measures $\{\tilde{O}_i\}$ which distinguish between non-omitters and omitters:

$$\tilde{O}_i = \begin{cases} 0 & \text{for} \quad O_i = 0 \\ 1 & \text{for} \quad O_i > 0 \end{cases}, \quad (i = 1, 2).$$

In addition, we define a measure M to distinguish between non-multiple omitters and multiple omitters:

$$M = \begin{cases} 0 & \text{for} \quad O_1 \leq 1 \\ 1 & \text{for} \quad O_1 > 1 \end{cases}.$$

Unlike omitting, guessing requires the construction of an operational definition. Angoff and Schrader [5] have noted that it is unlikely one can develop a satisfactory index of guessing based solely on examinee responses because of the effects of partial and incorrect knowledge possessed by examinees. To get a first order approximation, we assume, as in formula scoring, that all wrong answers are the result of random guessing. With this assumption, Ziller [42] offered an index of guessing based on taking the ratio of the estimated number of guesses, $w + \dfrac{w}{k-1}$, to the estimated number of guessing opportunities (case $i = 1$):

$$G_i = \frac{\dfrac{k}{k-1} w}{\dfrac{k}{k-1} w + O_i}, \quad (i = 1, 2),$$

where k is the number of choices per question and w is the number of incorrect responses by an examinee. The case $i = 2$, which truncates omissions after the last question answered, is a modification suggested by Angoff and Schrader [5]; it assumes that none of the trailing omitted questions were reached.

Rewriting G_i as

$$G_i = \left(1 + \frac{k-1}{k} \cdot \frac{O_i}{w}\right)^{-1}$$

observe that the guessing measures tend to underestimate guessing for test takers with partial knowledge because they are apt to get fewer wrong than random chance alone would explain, while the guessing measures tend to overestimate guessing for test takers with misinformation because they are apt to get more wrong than random chance alone would explain. This is not altogether unreasonable, since an informed guesser possesses more knowledge than an uninformed one and a misinformed guesser in effect possesses less knowledge. To some extent, under- and overestimation are a function of the quality of the response choices offered to examinees. Although these guessing measures are inherently imperfect, they are used by other researchers [5, 6], partly as rough indicators of guessing tendencies and partly for their value as collateral evidence of omitting tendencies.

G_1 and G_2 are undefined for the 16 examinees in our sample with perfect scores since we have no data for how often, if ever, they may have guessed. For calculations involving O_2 and G_2 about 10% of the raw data was not readily available, resulting in correspondingly reduced samples when processing these measures.

Not finishing the test may contribute to, or even account for, differential patterns in both guessing and omitting. This is of particular interest on tests that involve speed, as was the case with the test studied here, or when the test language is not the examinees' first language. To study differential patterns associated with not finishing, we utilize a dichotomous measure, NF, that distinguishes between those who answer at least one of the last two problems and those who do not:

$$NF = \begin{cases} 0 & \text{for} \quad O_1 - O_2 < 2 \\ 1 & \text{for} \quad O_1 - O_2 \geq 2 \end{cases}.$$

TABLE 2. Differential patterns in omitting and guessing

	% of omitters within subgroup			Estimated probability of guessing	
	$\widetilde{O}_1$	M	$\widetilde{O}_2$	G_1	G_2
Ethnicity (N)	(943)	(943)	(854)	(927)	(838)
Asian	10.6	6.3	6.6	.9725	.9847
Black	29.4	20.0	24.0	.9268	.9417
Latino	33.7	24.1	24.4	.9118	.9522
White	18.1	10.3	12.3	.9596	.9748
Language (N)	(748)	(748)	(695)	(735)	(682)
Chinese	6.7	2.3	2.3	.9904	.9966
English	22.7	15.3	16.7	.9425	.9620
French	40.6	25.0	35.7	.9159	.9367
Spanish	29.9	21.9	20.8	.9153	.9578
Age (N)	(1003)	(1003)	(908)	(987)	(892)
≤ 20	18.2	11.4	13.9	.9605	.9710
$21 - 24$	22.9	14.8	16.2	.9378	.9622
≥ 25	27.6	19.0	18.9	.9267	.9582
Gender (N)	(1003)	(1003)	(908)	(987)	(892)
Female	23.1	16.2	16.2	.9389	.9627
Male	20.2	10.8	15.5	.9568	.9699
Birthplace (N)	(943)	(943)	(894)	(927)	(858)
USA	24.2	16.1	18.6	.9347	.9556
Foreign	20.2	12.5	13.4	.9565	.9762

Results

Table 2 exhibits variation in omitting and guessing tendencies across sub-groups of the five demographic variables studied. Statistical analysis of the patterns shown in Table 2 is given in Tables 3A and 3B. These latter tables address the issues from three perspectives: (1) Is the variation statistically significant?; (2) For a given demographic variable, is the apparent variation or lack thereof attributable to variation in the other four variables?; (3) Is variation due to, or masked by, differences in prior mathematics skill across the levels of a given demographic variable? Table 3A presents the analysis for the three measures of omitting; the analysis for the two measures of guessing is given in Table 3B [1, 24]. Regression residual and influence diagnostics ([35]; PROC REG) suggested that it was inappropriate to use traditional regression analysis for the guessing measures G_1 and G_2; consequently, results reported in Table 3B were carried out with nonparametric procedures.

TABLE 3A. P-Values for tests of
within-group homogeneity of omitting

Group	$\widetilde{O}_1$			M			$\widetilde{O}_2$		
	Unadj.[a]	Adj. for Groups[b]	Adj. for Skill[c]	Unadj.[a]	Adj. for Groups[b]	Adj. for Skill[c]	Unadj.[a]	Adj. for Groups[b]	Adj. for Skill[c]
Ethnicity	.0001	.0005	.0001	.0001	.002	.0001	.0001	.0003	.0001
Language	.0001	.189	.002	.0004	.203	.005	.0001	.010	.001
Age	.011	.006	.077	.019	.002	.097	.235	.016	.497
Gender	.303	.141	.675	.023	.010	.274	.784	.238	.409
Birthplace	.141	.079	.006	.117	.110	.010	.034	.004	.002

[a] Cochran-Mantel-Haenszel test for variation in percentage of omitters across subgroups [35]; CMH2 in PROC FREQ.
[b] Controlling for the other four demographic variables ([35]; CMH2 in PROC FREQ). For $\widetilde{O}_1$ and M, $N = 704$; for $\widetilde{O}_2$, $N = 653$.
[c] Controlling for prior mathematics skill ([35]; CMH2 in PROC FREQ).

Ethnic variation in all five measures was statistically significant, even when controlling for the other demographic factors or for prior mathematics skill. Although the language variable also appeared to exhibit variation in all five measures, upon controlling for the other demographic factors, statistically significant variation remained only for the truncated measures, $\widetilde{O}_2$ and G_2. This suggested that language minority status was associated with omitting up to the time of stopping.

Age appeared to possess significant variation only in the nontruncated measures $\widetilde{O}_1$, M and G_1, with scoring advantage associated with youth (≤ 20 years of age). But, after controlling for the other demographic variables, there was some evidence of variation in the truncated measures as well. Prior mathematics skill (which was higher among youth, from Table 1) seemed more effective at explaining age variation—especially in guessing—than did the other four demographic factors.

TABLE 3B. P-Values for tests of
Within-Group Homogeneity of Guessing

Group	G_1 Unadj.[a]	Adj. for groups[b]	Adj. for skill[c]	G_2 Unadj.[a]	Adj. for groups[b]	Adj. for skill[c]
Ethnicity	.0001	.003	.0001	.0001	.003	.0001
Language	.0001	.186	.004	.0002	.003	.001
Age	.006	.055	.203	.217	.064	.777
Gender	.233	.111	.963	.757	.064	.732
Birthplace	.073	.036	.001	.017	.003	.0006

[a]Using ([35]; CMH2 in PROC FREQ) with G_1 and G_2 scores replaced by respective ranks.

[b]Controlling for the other four demographic variables ([35]; CMH2 in PROC FREQ). For G_1, $N = 691$; for G_2, $N = 640$.

[c]Controlling for prior mathematics skill ([35]; CMH2 in PROC FREQ).

Gender variation was negligible, except for women being more likely to be multiple omitters. This distinction, however, appeared to be explained by women having, on average, lower prior mathematics skill (Table 1).

Foreign and native (USA) born students differed in terms of the truncated measures; that is, before foreign students had run out of time, or otherwise stopped, they were more likely to guess and less likely to omit than were native born students. When controlling for prior mathematics skill (native born students scored higher, Table 1), differences were seen to extend to the nontruncated measures as well.

Prior mathematics skill not only uncovered hidden variation (in the case of birthplace), it also helped explain some of the apparent variation (age and gender). Due to its evident relevance, we investigated this variable in terms of its global relationships to guessing and omitting. In Table 4, we see that higher prior mathematics skill was associated with lower omitting and higher guessing probabilities [1, 25]. However, some of this association, especially with multiple omissions (M) and guessing before stopping (G_2), appeared attributable to demographic factors.

TABLE 4. Rank correlations of prior mathematics
skill with measures of omitting and guessing

	$\widetilde{O}_1$	M	$\widetilde{O}_2$	G_1	G_2
Rank Correlation	−.176	−.138	−.140	.154	.122
P-Value[a] (N)	.0001 (953)	.0001 (953)	.0001 (863)	.0001 (939)	.0006 (849)
P-Value[b] (N)	.022 (664)	.297 (664)	.087 (615)	.054 (659)	.121 (604)

[a]Cochran-Mantel-Haenszel test for rank correlation ([35]; CMH1 in PROC FREQ).

[b]Controlling for the five demographic variables.

To round out our picture of those being put at a disadvantage by number correct scoring, we asked: Among omitters, which subgroups are the ones most disadvantaged? Table 5 summarizes an analysis of the number of omissions made by an omitter. Only omitters for whom full demographic data was available ($N = 164$) were studied. The number of omissions was analyzed in two 5-way analysis-of-variance (ANOVA) models, one with and one without prior mathematics skill as a covariate. No significant variation was found across ethnic, first language, or birthplace groups. As reported in Table 5, males and those aged 21 and over tended to leave out more answers than did other omitters.

TABLE 5. Mean number of omissions by
omitters across age and gender subgroups

	Subgroup	N	Raw mean	Controlling for groups[a]		Controlling for groups and skill[b]	
				Adj. mean	P-value	Adj. mean	P-value
Age	≤ 20	59	2.76	2.26	.016[c]	2.18	.013[c]
	$21 - 24$	46	3.89	3.49		3.52	
	≥ 25	59	3.71	3.55		3.63	
Gender	Female	125	2.85	2.58	.036[c]	2.57	.039[c]
	Male	39	3.60	3.62		3.65	

[a] Fitting the additive ANOVA model with demographic group variables age, gender, ethnicity, language and birthplace ([35]; PROC GLM).

[b] Adding prior mathematics skill as a covariate to the model described in footnote [a].

[c] P-values for the F-test of equality of means, controlling for all other variables (Type III F-test).

Note: In both models, the group variables ethnicity, language, birthplace and skill each had P-values $> .3$.

To help clarify guessing and omitting tendencies, we studied the role played by not finishing the test. As displayed in Table 6, gender's relationship to not finishing varies with age and ethnicity. Not finishing is related to age (χ_2^2, $P = .049$); more specifically, the tendency to not finish increased with age ([35]; CMH1 in PROC FREQ, $P = .019$). This pattern remained after controlling for gender (CMH1, $P = .009$), prior mathematics skill (CMH1, $P = .032$) and gender and mathematics skill together (CMH1, $P = .020$). In terms of gender alone, not finishing appeared somewhat related (CMH1, $P = .055$), but after controlling for age the relationship became clearer (CMH1, $P = .027$) with the greatest gender difference among those aged 21-25. Observations in Table 6 also suggested that the women became frequent nonfinishers at an earlier age than did the men.

Ethnicity is strongly related to not finishing ($\chi_3^2 = 26.07$, $P = .0001$), but the pattern of variation may be seen in Table 6 to be gender related (Breslow-Day test for variation across ethnic groups in not finishing's association with gender yielded $P = .002$). Controlling for gender, ethnicity remained strongly associated

TABLE 6. Gender differences in not finishing

	Subgroup	N	Female (%)	Male (%)	P-value[a]
Age	≤ 20	399	4.65	2.04	.377
	$21-24$	270	9.73	2.35	.043
	≥ 25	264	9.15	7.00	.649
Ethnicity	Asian	197	2.11	0	.561
	Black	243	5.85	10.91	.228
	Latino	183	18.94	0	.0002
	White	256	3.95	4.81	.762

[a]Fisher's exact two-sided test.

with not finishing (Cochran-Mantel-Haenszel test for general association yielded $\chi_3^2 = 25.22$, $P = .0001$).

In terms of prior mathematics skill, F-tests in 1-way ANOVA showed that males in our study had higher scores than females among whites ($P = .008$), Asians ($P = .052$) and blacks ($P = .058$), while Latino males had an insignificant advantage ($P = .225$) over Latino females. Controlling for prior mathematics skill indicated a possible gender difference in not finishing probabilities among blacks (CMH2, $P = .061$). When controlling for prior mathematics skill for other ethnic groups, results were consistent with those seen in Table 6.

Finally, not finishing is associated with first language (χ_3^2, $P = .002$; and controlling for prior mathematics skill, CMH2, $P = .025$). Spanish speakers were more likely to not finish, and Chinese speakers were less likely, than were English and French speakers. However, controlling for ethnicity seemed to explain this variation across language (Cochran-Mantel-Haenszel for general association gave χ_3^2, $P = .939$). Especially noteworthy is that within the Latino subgroup, comparing the 45 Spanish speakers and 122 English speakers, there was no difference in not finishing ($P = .920$).

The change in a student's percentile score (or, equivalently, rank) due to formula scoring gives another indication of the degree to which number correct scoring can distort comparisons between omitters and non-omitters. The reranking, using formula scoring, raised the percentile score of the average omitter 3.28 percentile points. Looking, for example, at the 50th percentile, the effect of formula scoring among omitters ($N = 223$ of 1003) was to increase the number of omitters in the upper half by 42%, from 43 to 61. For multiple omitters ($N = 146$), formula scoring raised the percentile score by an average of 4.35 percentile points. Also, the number of multiple omitters above the 50th percentile increased by 68%, from 22 to 37.

Research Summary

Evidence has been presented showing differential patterns of guessing and omitting on a standardized mathematics placement test by ethnicity, first language, age, gender and birthplace. While prior mathematics skill seemed to account for variation across gender and age groups, it did not explain these differential patterns within the cultural groups, where blacks, Latinos, those born in the USA, and non-Chinese speakers all tended to be relatively more omissive and less likely to guess. Further, when group differences in not finishing the test were studied, evidence of differential patterns by gender were again uncovered, with (a) females in the age category 21-25 more likely to not finish than males and with (b) some mixed directions in gender tendencies depending on ethnicity: Latino females were more likely than Latino males to not finish and black males were somewhat more likely than black females to not finish. Here, gender differences in not finishing were not explained by prior mathematics skill. As a consequence, certain cultural and gender groups seemed disadvantaged by number correct scoring for reasons not entirely mathematical in nature. Formula scoring was found to counter the non-mathematical but negative performance effects, increasing the ranking of a typical omitter's score by over three percentile points.

Implications for Practice

Why do some examinees omit? Grandy [16] surveyed omitters on the GRE general test, shortly after a changeover by ETS from formula scoring to number correct scoring. Grandy's study indicated that a sizable number of omitters did not understand the instructions or chose not to follow them. Some claimed to have both read in the accompanying brochure and heard from test administrators the instruction not to guess, but these claims were not supported by the evidence. Others believed that random guessing helped only if you could eliminate some of the choices. Many did not seem to understand guessing strategies. Still others did not leave adequate time at the end to answer the remaining questions. A few expressed the belief that guessing would be a waste of time. Kahneman and Tversky [19] have suggested there may be personality issues that cause certain examinees to prefer a sure thing (not to be wrong) over a gamble (a possible wrong answer).

Although only modest numbers of students are near enough to any given cutoff score to have their placements affected, safeguarding fairness in testing is an overriding concern, especially at institutions with culturally diverse populations. In this context, the question of how best to reform placement testing needs to be addressed.

As a remedy, formula scoring with the correction for omitting, which was used in this study, eliminates the unfairness caused by omitting. However, most educators wince at the prospect of "padding" student's scores when there is no work on the part of students to justify it. An equally effective alternative to correct the unfairness caused by omitting is formula scoring with the correction for guessing. However, the pitfalls of more complicated instructions, increased test taking time, and an intensified emphasis on strategizing make this cure somewhat less attractive.

Without doubt, it is impossible to guarantee total fairness in scoring with traditional number correct scoring unless students answer all questions. Yet, it may still be possible to strengthen the procedures of test administration to resolve satisfactorily the observed unfairness in scoring. In this connection, we note the recent research of Budescu and Bar-Hillel [9], who endorsed number correct scoring over formula scoring on the basis of decision theoretic considerations, suggesting that students be admonished to "answer, answer, answer." This is consistent with Angoff and Schrader [5] who earlier made the point that determined, systematic efforts should be made to get students to answer all questions.

In a practical sense, some of the responsibility for fairness in number correct scoring is tacitly in student hands. Building on the research of Budescu and Bar-Hillel as well as Angoff and Schrader, we propose revising directions to examinees and explicitly sharing with them the responsibility of guaranteeing fairness in scoring. This might be accomplished, in part, by engaging test takers directly with a categorical statement such as:

> Because of the way this test is scored and interpreted, **to be fair to yourself you must answer all questions**, even if you must guess. You will not be penalized for guessing or incorrect answers.

Directions need to be framed broadly to appeal to examinees on multiple levels, not just the cognitive decision-theoretic level, to help students from diverse cultural and educational backgrounds make sense out of what is required.

When number correct scoring is used, a concerted effort needs to be put in place to inform students of their responsibility to insure fairness in their own score. This includes putting the above "truth in testing" consumer-like information in handouts and on test booklets, having test administrators explicitly repeat it and, to bolster the efficacy of this remedy, having test administrators alert examinees of the impending end of the test. For example, test administrators might announce the time, say, 1 or 2 or even 5 minutes before time is called, to help students fulfill their mandate to answer all questions. Although this approach will never totally eradicate omissions, it advances the interests of fairness for test takers and, to the extent that it succeeds in getting students to fill in all blanks, it serves the interests of test givers with improved comparability of scores, thereby reducing the likelihood that cultural and gender groups are disadvantaged due to scoring error.

REFERENCES

1. Agresti, A., *Categorical Data Analysis*, Wiley & Sons, New York, 1990.
2. Albanese, M. A., *The correction for guessing: A further analysis of Angoff and Schrader*, Journal of Educational Measurement **23** (1986), 225–235.

3. Albanese, M. A., *The projected impact of the correction for guessing on individual scores*, Journal of Educational Measurement **25** (1988), 225–235.

4. Angoff, W. H., *Does guessing really help?*, Journal of Educational Measurement **26(4)** (1989), 323–336.

5. Angoff, W. H., and Schrader, W. B., *A study of alternative methods for equating right scores to formula scores*, (ETS RR-81-8) (1981), Educational Testing Service, Princeton, NJ.

6. Ben-Shakhar, G., and Sinai, Y., *Gender differences in multiple-choice tests: The role of differential guessing tendencies*, Journal of Educational Measurement **28** (1991), 23–35.

7. Ben-Simon, A., *Psychometric and Cognitive Aspects of Partial Knowledge in Solving Multiple-Choice Tests*, (unpublished doctoral dissertation), The Hebrew University, Jerusalem, 1992.

8. Bridgeman, B. and Wendler, C., *Gender differences in predictors of college mathematics performance and in college mathematics course grades*, Journal of Educational Psychology **83(2)** (1991), 275–284.

9. Budescu, D. and Bar-Hillel, M., *To guess or not to guess: A decision-theoretic view of formula scoring*, Journal of Educational Measurement **30(4)** (1993), 277–291.

10. Burton, N. W., Lewis, C., and Robertson, N., *Sex Differences in SAT Scores*, (CEEB RR-88-9), College Entrance Examination Board, New York, 1988.

11. College Entrance Examination Board, *Elementary Algebra Skills*, a placement test, Descriptive Tests of Mathematics Skills, Educational Testing Service, Princeton, NJ, 1988.

12. College Entrance Examination Board, *Guide to the Use of the Descriptive Tests of Mathematical Skills,*, Educational Testing Service, Princeton, NJ, 1989.

13. *Comprehensive Tests of Basic Skills*, Mathematical Concepts and Applications, Form *U* Level *J* Mathematics, CTB/McGraw Hill, Monterey, CA, 1981.

14. Daunis, M. R., *The validity of placement testing in freshman mathematics at the University of Tennessee, Knoxville*, Dissertation Abstracts International **50/07-A 1914** (1988).

15. Fairtest, *Selected Annotated Bibliography on Language Minority Assessment*, National Center for Fair and Open Testing, Cambridge, MA, 1993.

16. Grandy, J., *Characteristics of Examinees Who Leave Questions Unanswered on the GRE General Test Under Rights-Only Scoring*, (ETS RR-87-38),, Educational Testing Service, Princeton, NJ, 1987.

17. Hudson, J. B., *An Analysis of ACT Scores, Placement Tests, and Academic Performance in Reading, English, and Mathematics Courses*, research report, University of Louisville, Kentucky, 1989.

18. Hyde, J. S., Fennema, E., and Lamon, S. J., *Gender differences in mathematics performance: a meta-analysis*, Psychological Bulletin **107(2)** (1990), 135–155.

19. Kahneman, D. and Tversky, A., *Prospect theory: An analysis of decision under risk*, Econometrica **47** (1979), 263–291.

20. Kanter, M. J., *An exploratory study of the relationship of demographic, Institutional, and assessment factors affecting access to higher education for underrepresented students in the California community colleges*, Dissertation Abstracts International **50/11-A 3455** (1989).

21. Klein, C. A., *Relationships between selected entry placement factors and performance on the college level academic skills test (Academic Skills Testing)*, Dissertation Abstracts International **53/10-A 3454** (1992).

22. Leder, G. C., *Mathematics and gender: changing perspectives*, Handbook of Research on Mathematics Teaching and Learning (D. A. Grouws, ed.), Macmillan, New York, 1992, pp. 597–622.

23. Mannan, G., Charleston, L., and Saghafi, B., *A comparison of the academic performance of Black and White freshman students on an urban commuter campus*, Journal of Negro Education **55(2)** (1986), 155–161.

24. Mantel, N., and Haenszel, W., *Statistical aspects of the analysis of data from retrospective studies of disease*, Journal of the National Cancer Institute **22** (1957), 719–748.

25. Mantel, N., *Chi-square tests with one degree of freedom: Extensions of the Mantel-Haenszel procedure*, Journal of the American Statistical Association **58** (1963), 690–700.

26. Mathematical Association of America, *Mathematical Association of America Placement Test Newsletter; v1-9, 1978-87* (Harvey, J. G., ed.), Mathematical Association of America, Washington, DC, 1987, pp. 1–148.

27. Mathematical Association of America, *A word about PTP and COT*, Mathematical Association of America Placement Test Newsletter, Winter (1991, 1992), 1.

28. Mathematical Association of America, *User's Guide: The Placement Test Program of The Mathematical Association of America*, Fourth Edition, Washington, DC, 1993.

29. Mathematical Sciences Education Board of the National Research Council, *Measuring What Counts: A Conceptual Guide for Mathematics Assessment*, National Academy Press, Washington, DC, 1993.

30. Meeks, K. I., *A comparison of adult versus traditional age mathematics students and the development of equations for the prediction of student success in developmental mathematics at the University of Tennessee–Chattanooga*, Dissertation Abstracts International **51/03-A 776** (1989).

31. Mislevy, R. J. and Verhelst, N., *Modeling item responses when different subjects employ different solution strategies*, Psychometrika **55** (1990), 195–215.

32. Parker, C. S. M., *Dispositional optimism and academic achievement on a required mathematics placement examination*, Dissertation Abstracts International **53/07-A 2207** (1992).

33. Reilly, R. R., *Empirical option weighting with a correction for guessing*, Journal of Psychological Measurement **35** (1975), 613–619.

34. Rosser, P., *The SAT gender gap: Identifying the causes*, Center for Women Policy Studies, Washington, DC, 1989.

35. SAS Institute Inc., *SAS/STAT User's Guide*, Version 6, Fourth Edition, vol. 1, 2, SAS Institute, Cary, NC, 1989.

36. Secada, W. G., *Race, ethnicity, social class, language, and achievement in mathematics*, Handbook of Research on Mathematics Teaching and Learning (D. A. Grouws, ed.), Macmillan, New York, 1992, pp. 623–660.

37. Slakter, M. J., *The penalty for not guessing*, Journal of Educational Measurement **5** (1968), 141–144.

38. Slakter, M. J., *The effect of guessing strategy on objective test scores*, Journal of Educational Measurement **5** (1968), 217–221.

39. Suciu, L. K., *Prediction of success in introductory college mathematics courses at Trident Technical College*, Dissertation Abstracts International **53/03-A 748** (1991).

40. Thorndike, R. L. (Editor), *Educational Measurement*, 2nd Edition, American Council on Education, Washington, DC, 1971.

41. Wilder, G. Z. and Powell, K., *Sex Differences in Test Performance: A Survey of Literature*, (CEEB RR-89-3), College Entrance Examination Board, New York, 1989.

42. Ziller, R. C., *A measure of gambling response-set in objective tests*, Psychometrika **22** (1957), 289–292.

DEPARTMENT OF MATHEMATICS AND STATISTICS, HUNTER COLLEGE OF THE CITY UNIVERSITY OF NEW YORK

CBMS Issues in Mathematics Education
Volume **6**, 1996

A Perspective on Mathematical Problem-Solving Expertise Based on the Performances of Male Ph.D. Mathematicians

THOMAS C. DEFRANCO

"The major part of every meaningful life is the solution of problems: a considerable part of the professional life of technicians, engineers and scientists, etc. is the solution of mathematical problems." (Halmos [6, p. 523])

As noted by mathematician and expositor Paul Halmos, learning to solve problems is an essential tool for life and has become an important component of the K-16 mathematics curriculum. Over the years mathematicians, psychologists, and educators have embraced various theories on problem solving in order to understand ways to teach students to become better problem solvers. In order to accomplish this goal researchers have examined differences between expert and novice problem solvers with respect to problem-solving behavior on various tasks.

Problem Solving Expertise

What constitutes problem solving expertise? The answer to this question has been evolving over the past 25 years. During the formative stages, information processing served as a framework in understanding how humans solve problems. This process, based on production rules and computer simulations, identified patterns of expert and novice problem solving behavior on well-defined problems. (For a comprehensive description of information processing theory see Newell and Simon [7].) As a result, the term "expert" became synonymous with someone who has accumulated a substantial amount of knowledge in a particular domain,

The author wishes to thank Stephen S. Willoughby, Peter Hilton, and Henry O. Pollak for their thoughts and guidance on the initial draft of this article. In addition, I am grateful to Alan H. Schoenfeld for his insightful comments and advice during the final draft of the manuscript.

that is, a content knowledge specialist. Schoenfeld [21] believed that the domain specific definition of expert appeared narrow and limiting since it eliminated the clumsy and unstructured problem-solving performances of individuals struggling and making progress while solving problems in unfamiliar domains.

In a series of research studies and summary articles Schoenfeld [21, 23] recast the definition of expert. His research indicated that problem-solving experts possessed a wide range of attributes that include: domain knowledge (the tools a problem solver brings to bear on a problem), problem-solving strategies (Polya-like heuristics), metacognitive skills (issues of control—selecting strategies and solution paths to explore or abandon, appropriate allocation of one's resources, etc.), and a certain set of beliefs (a particular world view of mathematics).

A pilot study was conducted to replicate Schoenfeld's work on problem-solving expertise. The eight subjects in the study (six males, two females) were Ph.D. mathematicians. As was common at the time, the study presumed expertise on the part of the experts—most people would consider a Ph.D. an expert mathematician, especially with regard to problems that had rather elementary content demands. Seven subjects were university faculty while the other taught high school. The subjects were asked to think aloud while solving four problems. The problems chosen for the study required a background in high school mathematics and perseverance and creativity in order to solve them. The results of the pilot study were quite disturbing. In many cases the experts could not solve the problems and exhibited few of the skills Schoenfeld attributed to experts. Why? How could this be?

The data indicated that the pilot study experts may not have been experts at all, that is, acquiring a Ph.D. in mathematics (a domain knowledge expert) does not guarantee an individual is an expert in solving problems. To remedy this situation a new study was conducted involving two groups of Ph.D. mathematicians.

The Study

Subjects.

The qualitative nature of this study demanded a small number of subjects be examined in depth. In addition, in order to control for gender differences, all males were selected to participate. Two groups were formed: a) group A—eight men who earned a doctorate or its equivalent in mathematics and have achieved national or international recognition in the mathematics community, and b) group B—eight men who earned a doctorate in mathematics but have not achieved such recognition. In the end, the original 16 contacted agreed to participate in the study.

Subjects in group A were chosen by professional and peer recommendations and according to the medals and honors awarded then by the mathematics community. Subjects in group B were chosen by professional and peer recommendations only. At the time of the study, 12 participants worked in an academic institution, three in industry, one was retired, and all lived in the Northeast. For more detailed information about the subjects (e.g. awards, medals, publications, etc.) see Appendix A.

Data collection.

Each subject was asked to solve four problems and to answer a questionnaire concerning his beliefs about mathematics and mathematical problem solving.

Problem-Solving Booklet. The problems chosen for the study were "ill- structured" problems, that is, no known algorithm or procedure is immediately available to the problem solver while solving the problem (Frederiksen [4]).

A search through problem-solving books (Ball [1]; Bryant [2]; Schoenfeld [15]; Polya [9, 10, 11]; Rapaport [12]; Shlarsky [24]; Trigg [26]; Wickelgren [27]) and journals (Schoenfeld [13, 14, 16, 18, 19]) produced a list of mathematics problems from which four were selected for the study. The problems can be found in Appendix B and the reader is invited to solve them.

Subjects were instructed to take as much time as necessary to solve each problem. In addition to writing down the solution, subjects were asked to think aloud while solving each problem. The researcher was present and took notes during the session. The notes were used to pose questions regarding the solution to the problem just completed. The entire session was audio taped. The audio tapes were transcribed, coded, and analyzed [17] and served as a means of understanding the solution process and the strategies used to solve each problem.

Mathematics Beliefs Questionnaire (MBQ). Beliefs, and in particular mathematics beliefs, play a significant role in an individual's effort and performance while solving problems (Schoenfeld [20, 21]; Silver [25]). A 10-item questionnaire was designed to elicit responses concerning an individual's beliefs about mathematics and mathematical problem solving. The MBQ can be found in Appendix C. The questionnaire and a blank tape were sent to each subject. Audio-taped responses were transcribed, coded and analyzed using procedures developed by the researcher [3].

Results and discussion.

The results of this study indicate clearly that the problem-solving performance of the group A mathematicians was that of experts (as defined by Schoenfeld [21, 23]) while the group B mathematicians performed like novices. (See Table 1 for a summary of the results.) Evidence will be demonstrated by examining the differences in the knowledge base, problem-solving strategies, metacognitive skills, and the beliefs of the mathematicians in groups A and B.

Knowledge base.

The knowledge base represents the "tools" a problem solver brings to bear on a problem. "It is intended as an inventory of all the facts, procedures, and skill—in short, the mathematical knowledge—that the individual is capable of bringing to bear on a particular problem" (Schoenfeld [21, p. 17]). Although all the subjects have acquired the necessary knowledge to solve the problems

TABLE 1. Summary of results

	Group A	Group B
Knowledge Base	Sufficient	Sufficient
Strategies	Productive	Minimal
Metacognitive Skills	Productive	Minimal
Beliefs	Productive	Counterproductive

a significant difference between the subjects in the two groups can be found in the depth and breadth of their research interests, and the number of articles, reviews, and books they've published. (For further information see Appendix A.) While solving the problems, in some instances, subjects in both groups could not recall the necessary mathematical facts/information needed to solve a problem in a particular way. In the end, lacking certain mathematical facts appeared to hinder the performance of the group B mathematicians while this did not effect the preformance of the group A mathematicians. For example, in responding to question 3 on the MBQ, subject B-2 stated,

> "OK, if I don't have enough information, facts or theorems whatever at my disposal, it may handicap me seriously and then I may not be able to make the right connections, that is, find similar problems. . . . For example, on the . . . first part of this interview there was a problem where I couldn't remember the law of cosines. Actually it was the law of sines that I should have remembered. I felt it was intimately related to the problem and yet I couldn't recall what it was or the exact structure of the result, even enough to kind of reconstruct it. So I couldn't recover the theorem and I felt that it was necessary to solve the problem. So to the extent that I was correct then a lack of this knowledge made the solution impossible."

Overall subjects in group B failed to solve or complete 12 (and possibly more) problems because they could not recall the necessary information needed to solve them.

Problem-solving strategies.

Each problem was rated and categorized as correct (correct approach/correct implementation) or incorrect (correct approach/incomplete implementation, correct approach/incorrect implementation, incorrect approach, no solution). In particular, problem 1 was rated correct if a subject reported 49 different ways to change one-half dollar (i.e., using quarters, dimes, nickels, and pennies). Problem 2 was rated correct if a response fell within the range of 10^{11} to 10^{16} cells. Solutions to problems 3 and 4 can be found in Polya [11] and Schoenfeld [17] respectively. These, along with other solutions generated by mathematicians not participating in the study, were used as a basis for rating problems 3 and 4. Otherwise a problem was rated incorrect and placed in one of the appropriate categories.

Of the 32 problems attempted by each group, subjects in group A solved 29 correctly while subjects in group B solved 7 correctly. (See Table 2.) In general, problems 2, 3, and 4 seemed unfamiliar to subjects in both groups. Of the 4 problems: problem 1 seemed easy but tedious for the subjects in group A while easy but unmanageable for subjects in group B; problem 2 was solvable for subjects in group A and difficult for subjects in group B; problem 3 seemed interesting and relatively challenging for subjects in group A while very difficult for subjects in group B; and problem 4 seemed very easy for subjects in group A and relatively difficult for subjects in group B. A brief summary of the approaches/strategies implemented by the subjects is discussed below.

TABLE 2. The number of correct and incorrect solutions
exhibited by subjects in both groups on problems 1–4

Group	Pr.	Correct	Incorrect			
			Correct approach/ incomplete	Correct approach/ incorrect	Incorrect approach	No solution
A						
	1	7	1			
	2	6		2		
	3	8				
	4	8				
Total		29	1	2		
B						
	1	2	3	2	1	
	2	1	4	2	1	
	3		4	1	1	2
	4	4	1	2	1	
Total		7	12	7	4	2

Problem 1: In how many ways can you change one-half dollar? (Note:
The way of changing is determined if it is known how many coins of
each kind—cents, nickels, dimes, quarters are used.)

Seven subjects in group A and two in group B solved this problem correctly by
fixing the number of quarters (2, 1, and 0) and then listing and enumerating the
remaining possibilities working from the larger to the smaller denominations.
Six subjects in group A stated other approaches (using generating functions,
equations involving integer solutions) to the problem but realized a simple listing-
and-counting approach was a more efficient way to proceed.

The remaining subjects in group B reported various approaches to the problem
(using generating functions, the inclusion-exclusion principle, integer solutions
to a Diophantine equation, enumerating all possibilities using 1 coin at a time,
using 2 coins at a time, using 3 coins at a time and finally 4 coins) but either
implemented the plan incorrectly or did not complete the solution.

Subjects in both groups were able to generate a variety of appropriate strategies to solve this problem. Since the answer is a relatively small number, the
most reasonable approach is to list a few cases, look for a pattern and count
the total number, that is a "brute force" approach. All subjects in group A had
prior experience solving this problem (all remembered seeing or doing this or a
similar problem before). Therefore, they had a ballpark estimate of the answer
and used this brute force method to solve the problem. In contrast, only two in
group B decided to use this method while the others used more sophisticated,
generalizable, unreasonable approaches (e.g., using generating functions to solve
this problem is analogous to using a canon to kill a fly) that ultimately led to an
incorrect solution. These results indicate that being able to choose a reasonable
approach to a problem from among a variety of viable approaches is an important
aspect in solving problems and maybe a trait of expert problem solvers. Further
research needs to be conducted on how one acquires an ability to distinguish
between reasonable and unreasonable approaches to a problem.

Problem 2: Estimate, as accurately as you can, how many cells might
be in an average-sized adult human body? What is a reasonable
upper estimate? A reasonable lower estimate? How much faith do
you have in your figures?

The solution to this problem requires little technical information but does
require the problem solver to make "good estimates" of average cell volume and
average body volume. Six subjects in group A and one subject in group B solved
this problem correctly. Of these subjects five in group A and one in group B
implemented a plan involving estimates of the volume of a cell and an average-
sized adult human body. The remaining subjects in group A implemented plans
involving estimates of the weight of a cell and an average-sized adult human body
(two subjects were correct) and estimating the number of atoms in a cell and
the weight of an average-sized adult human body. Of the six subjects in group

A who solved this problem correctly, five based their estimates of average cell volume on the magnification needed to see a cell through a microscope. Answers reported by subjects in group A were: 10^9, 10^{11}, 10^{12}, 10^{13}, 10^{14}, 10^{15}, 10^{16}, and 10^{20}.

The remaining subjects in group B offered various approaches to the problem—estimating the volume of a cell and an average-sized adult human body (three subjects), working either by weight or volume (one subject), working with the weights of the quantities (one subject), but in each case did not implement the plan. One subject estimated the volume of a cell and an average-sized adult human body but incorrectly estimated the volume of the cell. Another subject estimated the number of atoms in a cell and the volume of an average-sized adult human body while the remaining subject offered no plan or approach to the problem, reported some quick estimates and quickly (after 50 seconds) gave up on the problem. Answers given by subjects in group B were 10^{10}, 10^{13}, 10^{25}, while the remaining five subjects reported no solution.

> *Problem 3:* Prove the following proposition: If a side of a triangle is less than the average of the other two sides, then the opposite angle is less than the average of the other two angles.

Many of the subjects commented they could not recall ever seeing this problem before and some in group A were impressed with the simplicity and elegance of the proposition. All subjects in group A solved this problem correctly while no one in group B was able to do so. Six subjects (three in group A, three in group B) used the law of cosines while five subjects (three in group A, two in group B) used the law of sines to solve it. Of the remaining subjects in group A, one worked with a special case of the problem while the other used a string of trigonometric identities involving the sine and cosine of the angles.

Two of the three remaining subjects in group B used techniques from analytic geometry while the remain subject explored three different approaches before reporting that the statement of the problem was incorrect.

> *Problem 4:* You are given a fixed triangle T with base B. Show that it is always possible to construct, with ruler and compass, a straight line parallel to B such that the line divides T into two parts of equal area.

All subjects in group A and four subjects in group B solved this problem correctly. Seven subjects in group A implemented a plan which involved: a) recognizing the upper triangle is similar to the given triangle T, and b) calculating the distance from the upper vertex of T to the line parallel to the base of T such that the line divides T into two parts of equal area. The remaining subject in A began with a right triangle and proceeded in a way similar to the other subjects. Six subjects did the construction during the solution of the problem.

In group B, two subjects implemented a plan similar to the one used by the seven subjects in group A, one used Galois Theory and one used analytic geometry to solve this problem. The remaining subjects implemented plans involving: a) ideas from analytic geometry (one subject),b) the center of gravity of the figure (one subject), and c) the similarity property between the triangles (two subjects). In each case the subject did not complete the solution. Two subjects reported how to do the construction during the solution of the problem.

Metacognitive skills-control.

The term " ... metacognition has two separate but related aspects: a) *knowledge and beliefs* about cognitive phenomena, and b) the regulation and *control* of cognitive actions" (Garofalo [5, p. 163]). The regulation and control component of metacognition suggests an active monitoring system that helps keep a solution on a correct path. Control decisions in problem solving have a major impact on a solution path and actions taken at this level typically include decisions regarding: a) solution paths to explore (or not to explore), b) abandoning, pursuing, changing or selecting approaches/strategies on a problem and c) the allocation of resources at a problem solver's disposal (Schoenfeld [21, p. 27]).

After examining the problem-solving behavior of experts (Ph.D. mathematicians) on problems similar to the ones in this study, Schoenfeld [21] categorized four types of control decisions. The results of this study suggest that a fifth category (Type A) should be added. The five categories observed in the data are given in Table 3.

The following protocol presents the problem-solving performance of John (a pseudonym for subject A-1) on problem 3 and is an example of Type D control behavior.

Problem-solving protocol.

(1) Prove the following proposition: If a side of a triangle is less than the average of the other two sides, then the opposite angle is less than the average of the other two angles.

(2) The side of the triangle is less than the average, OK ...

(3) The second statement, the conclusion is equivalent to the opposite angle is less than $60°$ ...

(4) Because they ... they add up to $180°$ and if it's less than the average then it's less than $60°$ ...

(5) What can you get from that ...

(6) The law of cosines tells you ... $x^2 - xy(\cos 60°) + y^2 = z^2$ ...

(7) And this is really something ... less than $\cos 60°$ ...

(8) $\cos 60°$ is $1/2$ and there must be a 2 in there to make it come out right ... OK, so $x^2 - xy + y^2$ ...

(9) I don't think this is true ...

(10) $x^2 - xy + y^2$, let's try a few examples here ...

(11) So when, so when it's equal to the average, the easy case is 4, 3, 5 right triangles ...

(12) Four is the one in question ...

TABLE 3. The effects of different types of control decisions
on problem-solving success: a spectrum of impact

Type A.	There is (virtually) no need for control behavior. The appropriate facts and procedures for problem solution can not be accessed from long-term memory. For all practical purposes, the problem solver gives up and resources are not exploited at all.
Type B.	Bad decisions guarantee failure: Wild goose chases waste resources, and potentially useful directions are ignored.
Type C.	Executive behavior is neutral: Wild goose chases are curtailed before they cause disasters, but resources are not exploited as they might be.
Type D.	Control decisions are a positive force in a solution: Resources are chosen carefully and exploited or abandoned appropriately as a result of careful monitoring.
Type E.	There is (virtually) no need for control behavior: The appropriate facts and procedures for problem solution are accessed in long-term memory.

---

(13) And you ask is that less than $60°$ ...

(14) Well, it is ... OK ...

(15) So z^2 ...

(16) OK, the angle is less than $60°$ ...

(17) So the cosine is more than the $\cos 60°$ ...

(18) So you've got $x^2 - xy + y^2$ ...

(19) And that falls with one relation or another to ...

(20) z^2 ... I'll fill in the greater than or less than sign later to see which one it's supposed to be ...

(21) z^2, now what we're concerned about is z^2, z ... compared to the average is $(x + y)/2$ ...

(22) So we're clearing the fractions, we got $4x^2 - 4xy + 4y^2$ versus $x^2 + 2xy + y^2$ ...

(23) Does that look right ...

(24) $3x^2 - 6xy + 3y^2$ which is $3(x - y)^2$ which is positive ...

(25) So now I believe it's true ...

(26) Because you've got something that's a square. Now, you've just got to work out the signs to make it come out right ...

(27) $3(x - y)^2$ is positive, that means that $4x^2 - 4xy + 4y^2 > (x + y)^2$ ... and that means that $x^2 - xy + y^2$ is bigger than ... bigger than or equal to

... the average squared ...

(28) This isn't working ...

(29) The average is less than ... that, which is equal to, that is $x^2 - xy + y^2$ ... which is equal to $x^2 - 2xy(\cos\theta) + y^2$ and now which way ... which one ...

(30) θ is less ... $\cos\theta$ is ... more ... $-2xy(\cos\theta)$ is less ...

(31) So that $x^2 - 2xy(\cos\theta)$ ... this isn't working ...

(32) Is less than some quantity that the average is also less than ...

(33) That doesn't make any sense ...

(34) If the side, if the side is less than the average of the sides ...

(35) Well, let me read the problem over ...

(36) If the side is less than the average of the sides then the angle is less than the average of the angles ...

(37) OK ... so, the side z is ... how embarrassing, I have to go to a second sheet on my own subject ...

(38) Now $z < (x + y)/2$, square both sides ...

(39) So that $4z^2$ is less than, not less than or equal to, less than $(x + y)^2$ ...

(40) And $4z^2$ is in turn $4x^2 - 4xy(\cos\theta) \ldots 8xy(\cos\theta) + 4y^2$ ...

(41) Expand both sides ... $4x^2 - 8xy(\cos\theta) + 4y^2 < x^2 + 2xy + y^2$ ...

(42) So in other words ... $3x^2 - 6xy + 3y^2$ ... I have to get the $(x - y)^2$ in there somehow and I'll work out the other things later ...

(43) Is less than $8xy(\cos\theta) - 4xy$ ...

(44) So ... and zero is less than the whole smear ...

(45) Because $0 < 3(x - y)^2 < 4xy(2\cos\theta - 1)$ ...

(46) So $2\cos\theta - 1$ is positive ...

(47) $\cos\theta > 1/2$ ...

(48) That tells us that $\theta < 60°$.

Problem-solving analysis.

John read the problem and identified the conditions and the goal of the problem (Statements: 1-2). He decided the law of cosines would be a plausible way to approach the problem (6) but realized inconsistencies in his work (9) and felt compelled to check whether the statement of the problem was true.

At this point, John decided to examine a special case of the problem (a Polya-like strategy), that is, a 3-4-5 right triangle. By examing and learning from this example (11-24), John was convinced the statement of the problem was true (25). Once again he returned to his original plan (the law of cosines) but soon recognized contradictions in his work (33). This led him to reread the problem (34).

Immediately he realized he misinterpreted the statement of the problem. Working from the conditions of the problem and using the law of cosines, he worked in an efficient and accurate manner and completed a correct solution to the problem (36-48).

This protocol is an example of control at its best and the role it can play in solving a problem. Schoenfeld [21] discovered that individuals exhibiting efficient control behavior maintain an internal dialogue and argue with themselves while solving a problem. " ... Plans are not made, they are evaluated and contrasted with other possible plans ... Solutions are monitored and assessed 'on line,' and signs of trouble suggest that current approaches might be terminated and others considered ... " (p. 140)

It is apparent that John monitored, assessed, and evaluated his work effectively throughout the entire process. In short, control became a positive force in helping him solve this problem. His performance is indicative of Schoenfeld's notion of what it means to be an expert problem solver–behavior that takes into account individuals who solve problems but navigate the solution space making progress in a meandering but meaningful manner.

Using Schoenfeld's work [21] on control as a guide the problem-solving protocols were analyzed and categorized into the various types of control behavior (see Table 4). An examination of Table 4 illustrates quite dramatically the difference between the subjects in the two groups regarding control behavior during the solution of the problems.

TABLE 4. Types of problem-solving behavior
exhibited by subjects in both groups on problems 1-4

	Type				
Groups	A	B	C	D	E
A			2	22	8
B	5	3	17	7	

Beliefs.

Schoenfeld [21] recognized that mathematical behavior, which appears to be solely cognitive in nature, may in fact be influenced by affective components (p. 155). For example, beliefs regarding problem solving (e.g., perseverance, confidence, motivation, interest, etc.) contribute significantly to an individual's performance on a problem. The questionnaire responses indicated clearly that the beliefs of subjects in group A are different than those in group B with respect to mathematics and problem solving. Responses to the questionnaire are briefly discussed next.

On question 1 (see Appendix C) subjects in both groups indicated the most important characteristics or qualities of an expert problem solver include: experience, knowledge of mathematics, the use of analogies, confidence, perseverance and motivation. In addition, subjects in group A believed successful problem-solving experiences contribute significantly to becoming an expert while subjects in group B seemed to place more emphasis on motivation and perseverance.

According to subject A-4,

> "First, I say a deep knowledge of mathematics itself ... and a real understanding of interrelations between the different parts of mathematics and of the relationships of mathematical models to real world problems ... Beyond that there is the issue of confidence based on past experience of success ... "

Subject B-3's response was representative of group B. He stated,

> " ... you have to be well motivated to do it, in that you think the problem is worth solving, that there is a value in ... in doing it in some sense ... Another thing that will make for decent problem solving ability is your depth in that field ... Third, it's pertinacity. It's related to the ... you've really got to stick to significant problems for many, many, many years. Somehow, not necessarily years but for a lot of time."

On question 2, 12 subjects (seven in group A, five in group B) noted the importance of a "good memory" in solving problems while 11 subjects (seven in group A, four in group B) understood the significance of recalling analogous problems for successful problem solving.

On the role memory plays in relation to problem solving subject A-4 stated,

> " ... One does not call upon facts, information, and theorems as if they are simply part of one's library. There is an interaction between the progress one makes in attempting to understand the problem and the recall of facts and information. But it is very much a matter of conducting an efficient search within a relatively small area for relevant facts, information, and theorems and that of having a vast stock of such facts available. It is by one's ability to localize the search that one renders oneself efficient as a problem solver."

According to subject B-2,

> "OK, if I don't have enough information, facts or theorems whatever at my disposal, it may handicap me seriously and then I may not be able to make the right connections, that is, find similar problems ... to ... get a handle on solving the problem ... "

Subject A-3 theorized several kinds of memory may exist—a "memory for analogies or associations" and a "memory for discrete facts" and indicated both are necessary while solving problems. He stated,

"It's my feeling there are several kinds of memory. One is the memory of a specific theorem, a result that might be useful in solving the problem or a specific technique. But there may be a deeper, intuitive memory in which you see a problem of a certain type and you somehow feel that some approaches will be more successful than others. Even though you can't put your finger on any specific technique or theorem which will apply but you may recast the problem into ... something where you feel more comfortable or more likely to come up with the appropriate solution."

On question 3, subjects were asked to comment on the issue of not having enough knowledge to solve a problem. Nine subjects (four in group A, five in group B) responded they would not attempt the problem, while four subjects (three in group A, one in group B) stated they would use external sources in order to solve the problem. The remaining responses included that this fact would: 1) have no effect at all, and 2) lead to an incorrect solution to the problem.

While solving the problems subjects in group A exuded confidence and possessed a self-assured, almost cocky attitude in their ability. They were aware of their reputation and enjoyed the challenge of not only finding a correct solution but solving the problem as quickly as possible. The same attitude was echoed in their responses to question 4—seven out of eight considered themselves to be expert problem solvers. On explaining why he felt this way, subject A-4 stated,

" ... Yes I do consider myself to be fairly good at problem solving and that is because I believe, first, I do understand the role of mathematics in problem solving and, second, because I have had a sufficient history of success to make me rather confident in my ability to solve problems."

In contrast only one subject in group B believed he was an expert problem solver. The responses indicated a significant difference in the level of confidence between the groups.

Question 5 asked the subjects to describe general strategies that would be useful in solving a problem. The strategies cited as most helpful are recalling and using analogous problems (five in group A, three in group B) and examining individual or special cases of a problem (three in each group).

On the issue of using alternative methods to solve a problem (question 6), five subjects in group A indicated they almost always rework problems while the remaining three do so under certain conditions. Four subjects in group B stated almost never do they look for alternative ways to solve problems while only three stated almost always and one under certain conditions.

On this issue subject A-6 stated,

> "Well the answer is always. I never let anything go to bed. I never
> consider it done and always look for an alternative method ... [it isn't]
> quite always ... but usually I always go over and look for another
> solution and play with it. You see that's kind of a nice thing for
> me, at any rate because the painful job has been done and now I'm
> reaping the harvest. Playing with it and redoing it and seeing how it
> fits into other places is just fun, it's easy."

It seems, according to the phrase in the last sentence, "seeing how it fits
into other places," this subject has acquired the mathematician's aesthetic, that
is " ... a predilection to analyze and understand, to perceive structure and
structural relationships, to see how things fit together" (Schoenfeld [22, p. 87]).
According to Schoenfeld, developing this aesthetic is a core aspect of learning to
think mathematically. Arguably, the subjects in group B may not have acquired
this aesthetic.

Group A subjects stated more mathematical areas they enjoyed and usually
worked on than the subjects in group B. In addition, group A subjects stated
fewer mathematical areas they did not enjoy and usually did not work on than
the subjects in group B (questions 7 and 8).

Subjects in both groups believed that having confidence in an area of mathematics would enhance their ability to solve problems in that area (question 9).
Similarly, subjects in group A believed that a lack of confidence would not deter
them from solving the problem while subjects in group B felt a lack of confidence
would affect their ability to solve problems. Subject B-2's response is typical of
group B. He stated,

> "I think it would ... have a great effect. I think I would be demoral-
> ized. I would be disinclined to expend a serious effort on this problem.
> I would doubt my abilities to do so. I would give up the ghost, you
> know, very easily. I would ... be anxious to be rid of having to grap-
> ple with this problem and I'd quit as soon as possible. I may make a
> show of looking at it but I'd be pretty quick to throw up my hands."

In short, the responses to the questionnaire reveal that the beliefs about mathematics and problem solving held by subjects in group A are dissimilar to those
held by the subjects in group B. To the extent that beliefs impact problem solving
performance, it would appear the beliefs acquired by group A (group B) subjects
would positively (negatively) influence their performance on the problems.

Summary and Final Remarks

Expert-novice paradigms have served as a means of studying individual differences in various problem-solving contexts. (A brief overview of the expert-novice
research can be found in Owen [8].) As a result of the research in this area the
term expert came to be known as someone who knows the domain cold and can
solve problems in a nearly automatic and routine fashion. Schoenfeld [21] argued

that this type of behavior represented a certain type of competency and failed to reflect the notion that *problem solving experts*, as opposed to *content experts*, are people who manage to solve problems even when the solution is not readily apparent to them. He believed that in addition to knowing the domain, problem solving expertise involves other attributes such as, problem-solving skills, metacognitive skills, and a certain set of mathematical beliefs. A pilot study was conducted to replicate his work on expertise. The results were disappointing and indicated that the ostensible experts in the study, eight Ph.D. mathematicians, may not have been problem solving experts at all. To test this idea a new study was conducted involving two groups of male Ph.D. mathematicians. Group A consisted of eight Ph.D. mathematicians with a national or international reputation in the mathematics community while group B consisted of eight Ph.D. mathematicians without such a reputation.

In the study, subjects were asked to think aloud while solving four mathematics problems (Appendix B) and answer a questionnaire regarding their beliefs about mathematics and problem solving (Appendix C). In solving the problems, the group A mathematicians outperformed the group B mathematicians by a wide margin. Further analysis revealed differences between subjects in the two groups in: a) their knowledge base (see Appendix A), b) their problem-solving skills (in many instances, the strategies used by the group A subjects were helpful in solving the problems while those used by the group B subjects were less helpful and sometimes counterproductive), c) their metacognitive skills (while solving the problems control was a positive force for the group A subjects and control was either missing or a negative force for the group B subjects), and d) their beliefs about mathematics and problem solving (the mathematical belief system of subjects in group A was different than that of the subjects in group B).

Although all the mathematicians in group B possess a strong content knowledge base and have produced some research, they are not problem-solving experts. The results of this study indicate clearly that it is possible for people to be content experts while possessing only modest problem solving skills—that problem solving expertise is a property separable from content expertise. In addition, the evidence from this study begins to focus the picture of what it means to be a problem solving expert and lends strong support to Schoenfeld's research [21, 23] in this area.

The second implication of this study raises the difficult issue of how we train undergraduate and graduate mathematicians at our universities. It is apparent that university mathematics departments train students in subject matter but not in problem solving skills. To the extent that solving problems is important (see the opening quote) and to the extent that training students in problem-solving skills is possible (and I believe it is) then the mathematics community needs to rethink the culture in which students are trained to be mathematicians.

References

1. Ball, W. W. R., *Mathematical Recreations and Essays*, (revised edition), Macmillan, New York, 1962.

2. Bryant, S. J. et al., *Non-Routine Problems in Algebra, Geometry and Trigonometry*, McGraw-Hill, New York, 1965.

3. DeFranco, T. C., *The role of metacognition in relation to solving mathematics problems among Ph.D. mathematicians*, (Doctoral dissertation, New York University), 1987.

4. Frederiksen, N., *Implications of cognitive theory for instruction in problem solving*, Review of Educational Research **54(3)** (1984), 366–407.

5. Garofalo, J. and Lester, F. K., *Metacognition, cognitive monitoring and mathematical performance*, Journal for Research in Mathematics Education **16(3)** (1985), 163–176.

6. Halmos, P. R., *The heart of mathematics*, The American Mathematical Monthly **87(7)** (1980), 519–524.

7. Newell, A. and Simon, H. A., *Human Problem Solving*, Prentice-Hall, Englewood Cliffs, NJ, 1972.

8. Owen, E. and Sweller, J., *Should problem solving be used as a learning device in mathematics?*, Journal for Research in Mathematics Education **20(3)** (1989), 322–328.

9. Polya, G., *Mathematical Discovery*, (2 Vols.), John Wiley & Sons, New York, 1962, 1965.

10. Polya, G., *How to Solve it*, (2nd ed.), Princeton University Press, Princeton, NJ, 1973.

11. Polya, G. and Kilpatrick, J., *The Stanford Mathematics Problem Book*, Teachers College Press, New York, 1974.

12. Rapaport, E., *The Hungarian Problem Book*, (translated by E. Rapaport), Mathematical Association of America, Washington, DC, 1963.

13. Schoenfeld, A. H., *Explicit heuristic training as a variable in problem-solving performance*, Journal for Research in Mathematics Education **10** (1979), 173–187.

14. Schoenfeld, A. H., *Presenting a model of mathematical problem solving*, paper presented at the annual meeting of the AERA, San Francisco, CA (1979).

15. Schoenfeld, A. H., *Heuristics in the classroom*, Problem Solving in School Mathematics, 1980 Yearbook of the National Council of Teachers of Mathematics (S. Krulik, ed.), The National Council of Teachers of Mathematics, Reston, VA, 1980.

16. Schoenfeld, A. H., *Teaching problem solving skills*, American Mathematical Monthly **87(10)** (1980), 194–205.

17. Schoenfeld, A. H., *Episodes and executive decisions in mathematical problem solving*, paper presented at the annual meeting of the AERA, Los Angeles, CA, (April, 1981).

18. Schoenfeld, A. H., *Measures of problem-solving performances and of problem solving instruction*, Journal for Research in Mathematics Education **13(1)** (1982), 31–49.

19. Schoenfeld, A. H. and Hermann, D., *Problem perception and knowledge structure in expert and novice mathematical problem solvers*, Journal of Experimental Psychology: Learning, Memory and Cognition **8(5)** (1982), 484–494.

20. Schoenfeld, A. H., *Beyond the purely cognitive: belief systems, social cognitions and metacognitions as driving forces in intellectual performance*, Cognitive Science **7** (1983), 329–363.

21. Schoenfeld, A. H., *Mathematical Problem Solving*, Academic Press, Inc., 1985.

22. Schoenfeld, A. H., *Problem solving in context(s)*, The Teaching and Assessing of Mathematical Problem Solving (R. I. Charles and E. A. Silver, eds.), vol. 3, The National Council of Teachers of Mathematics, Reston, VA, 1989, pp. 82–92.

23. Schoenfeld, A. H., *Learning to think mathematically: Problem solving, metacognition, and sense making in mathematics*, Handbook of Research on Mathematics Teaching and Learning (D. A. Grouws, ed.), Macmillan Publishing Co., 1992, pp. 334–370.

24. Shlarsky, D. D. et al., *The USSR Olympiad Problem Book*, Freeman, San Francisco, CA, 1962.

25. Silver, E. A., *Knowledge organization and mathematical problem solving*, Mathematical Problem Solving: Issues in Research (F. K. Lester and J. Garofalo, eds.), Franklin Institute Press, Philadelphia, PA, 1982, pp. 15–25.

26. Trigg, C. W., *Mathematical Quickies*, McGraw-Hill, New York, 1967.

27. Wickelgren, W., *How to Solve Problems*, Freeman, San Francisco, CA, 1974.

Appendix A

Subjects	[a]Publications	
	[b]TA	[c]Reviews
A1	28	—
A2	43	38
A3	143	45
A4	274	169
A5	164	60
A6	135	—
A7	33	29
A8	16	190
Total	836	531
B1	9	15
B2	15	5
B3	33	100
B4	6	—
B5	10	—
B6	52	59
B7	4	—
B8	3	—
Total	132	179

Awards–Group A[d]: Honorary Degrees Awarded–12, Silver Medal (University Helsinki), Centenary Medal (John Carroll University), Polya Prize in Combinatorics, Newcomb Cleveland Prize, National Medal of Science, Wolf Prize, Ford Prize, Numerous Visiting Professor Appointments.

Elected Positions–Group A[d]: Co-Chairman, Cambridge Conference on School Mathematics; Chairman, US Comm. on Math Inst.; President–AMS; Vice President–AMS; President–MAA; Vice President–MAA; Member–National Science Board.

No distinguished awards or elected positions were found for subjects in group B.

[a]Available information found in the Math/Sci Database (1940–1992)

[b]TA=Total author-published articles, contributions to articles/books, editor of books

[c]Reviews—Reviews of articles

[d]Available information in American Men and Women in Science–1992

Appendix B

Problem 1

In how many ways can you change one-half dollar? (Note: The way of changing is determined if it is known how many of each kind–cents, nickels, dimes, quarters are used.)

Problem 2

Estimate, as accurately as you can, how many cells might be in an average-sized adult human body? What is a reasonable upper estimate? A reasonable lower estimate? How much faith do you have in your figures?

Problem 3

Prove the following proposition: If a side of a triangle is less than the average of the other two sides, then the opposite angle is less than the average of the two other angles.

Problem 4

You are given a fixed triangle T with base B. Show that it is always possible to construct, with ruler and compass, a straight line parallel to B such that the line divides T into two parts of equal area.

Appendix C

1. Please describe the qualities, characteristics or factors that you think make an individual an expert problem solver in mathematics.

2a. Suppose you are asked to solve a mathematics problem (i.e., either a research problem or a textbook problem and one that you do not recall doing before). How does your memory for facts, information, theorems, etc., affect your problem solving?

2b. What effect do you think this fact (i.e., your answer to part a) may have upon your ability to solve the problem?

2c. Why?

3a. Suppose you are asked to solve a mathematical problem and immediately after reading the problem you realize that you do not think you have enough knowledge to solve the problem. What effect do you think this fact might have upon your ability to solve the problem?

3b. Why?

4a. Do you consider yourself to be an expert problem solver in mathematics?

4b. Why?

5. Suppose you are asked to solve a mathematics problem (i.e., either a research problem or a textbook problem and one that you do not recall doing before). What general strategies or techniques do you think you would use to help you toward the solution of the problem?

6a. After solving a mathematics problem when do you rework and use or not use alternative methods to solve the problem?

6b. Why?

7. Please describe the type[s] of mathematics problem[s] you enjoy and usually work on.

8. Please describe the type[s] of mathematics problem[s] you do *not* enjoy and do *not* usually work on.

9a. Which areas or branches of mathematics do you feel most confident working in?

9b. Suppose that a mathematics problem you are working on falls in one of the areas or branches of mathematics you feel most confident working in. What effect do you think this would have upon your ability to solve the problem?

9c. Why?

10a. Which areas or branches of mathematics do you feel least confident working in?

10b. Suppose that a mathematics problem you are working on falls in one of the areas or branches of mathematics you feel least confident working in. What effect do you think this would have upon your ability to solve the problem?

10c. Why?

UNIVERSITY OF CONNECTICUT, COLLEGE OF EDUCATION

CBMS Issues in Mathematics Education
Volume **6**, 1996

Questions on New Trends in the Teaching and Learning of Mathematics

THE OBERWOLFACH CONFERENCE
27 NOVEMBER–1 DECEMBER, 1995

The following research, curricular, and pedagogical questions arose in response to the presentations given at this conference. They represent some of the important issues and problems that the participants jointly agree should be studied. The list is far from complete and should not be interpreted as an attempt to put forth a research program or agenda. Neither is it claimed that these questions are original or of equal importance. We only hope that they will serve to stimulate workers in the field to obtain new results and to improve the learning of mathematics by students throughout the world.

1. What are appropriate methodologies for answering curricular and pedagogical questions?

2. Are learning theories transferable across cultural and subject matter boundaries? Can they be applied to different topics and different groups of students in different countries?

3. What are the different learning styles for mathematics that are prevalent among post-secondary students? How do these learning styles relate to various theories of learning? How immutable is the learning style of an individual student?

4. What are the differences between how mathematics is learned by experts and by novices of different kinds?

5. What do faculty and students mean by the word "understanding"? What is meant by "clarity"? What is the relationship between clarity and precision in the minds of students and faculty?

6. Do the tools of technology change students' understanding of mathematics, and if so how? For example: some people argue that learning geometry with a software package does not promote the same understanding of geometry as learning in a paper and pencil environment. How can we transform this claim into a research question and what methodology can be developed to investigate this question?

7. What are the student conceptions of the different notions of equality and approximate equality? How are these conceptions affected by technology?

8. What are the difficulties that students have with formal mathematical language such as the use of "for all," "there exists," two-level quantifiers, and negation, and with the relationship of formal mathematical language to everyday language?

9. Why is the concept of a solution to a differential equation difficult? What is the nature of that difficulty? In particular, do students find it difficult to understand—symbolically, graphically, and visually—what it means to be a solution to a differential equation or initial value problem?

10. What pedagogical strategies can be effective in helping students understand the systematic development of mathematical theories?

11. How can we most effectively teach students to use definitions as a mathematician does, and in particular to turn a definition into "an operative form"?

12. What is the relationship between time spent on mathematics outside of class and the level of student understanding? What pedagogical strategies are most effective in improving the quantity and quality of the time students spend on mathematics?

13. What course designs and pedagogical strategies are most effective in taking into account the wide range of abilities and backgrounds of the students?

14. What are the pedagogical advantages and disadvantages of the different ways in which technology can be used? Among these are visualization, the use of built-in mathematical tools, and programming.

15. How does class size affect learning? How is this affected by technology and cooperative learning? What group sizes in cooperative learning best support learning?

16. What are the advantages and disadvantages of using applications from both inside and outside mathematics and of using history? Do they improve the students' retention of the mathematics and/or the retention of the students in mathematics? What is their effect on understanding, and the appreciation of mathematics both for its internal beauty and its usefulness?

17. What form or forms of proof are appropriate in different contexts for student learning and how should they be dealt with pedagogically?

18. What algebra is appropriate as preparation for post-secondary work? How is the answer affected by subject? How is it affected by technology?

Participants

David Bressoud
Urs Kirchgraber
Ed Packel
Bill Barker
Ed Dubinsky
Werner Hartmann
Lisa Hefendehl-Hebeker
Wolfgang Henn
Reinhard Hoelzl
Deborah Hughes-Hallett
Hans Niels Jahnke
Dan Kennedy
Heinz Klemenz
Colette Laborde
Jean-Marie Laborde
Hans-Christian Reichel
V. Frederick Rickey
Werner Schmidt
Inge Schwank
David Smith
Anita Solow
John Stillwell
David Tall
Bernd Wollring

OBERWOLFACH

27 NOVEMBER–1 DECEMBER, 1995

RESEARCH IN COLLEGIATE MATHEMATICS EDUCATION

EDITORIAL POLICY

The papers published in these volumes will serve both pure and applied purposes, contributing to the field of research in undergraduate mathematics education and informing the direct improvement of undergraduate mathematics instruction. The dual purposes imply dual but overlapping audiences and articles will vary in their relationship to these purposes. The best papers, however, will interest both audiences and serve both purposes.

CONTENT.

We invite papers reporting on research that addresses any and all aspects of undergraduate mathematics education. Research may focus on learning within particular mathematical domains. It may be concerned with more general cognitive processes such as problem solving, skill acquisition, conceptual development, mathematical creativity, cognitive styles, etc. Research reports may deal with issues associated with variations in teaching methods, classroom or laboratory contexts, or discourse patterns. More broadly, research may be concerned with institutional arrangements intended to support learning and teaching, e.g. curriculum design, assessment practices, or strategies for faculty development.

METHOD.

We expect and encourage a broad spectrum of research methods ranging from traditional statistically-oriented studies of populations, or even surveys, to close studies of individuals, both short and long term. Empirical studies may well be supplemented by historical, ethnographic, or theoretical analyses focusing directly on the educational matter at hand. Theoretical analyses may illuminate or otherwise organize empirically based work by the author or that of others, or perhaps give specific direction to future work. In all cases, we expect that published work will acknowledge and build upon that of others – not necessarily to agree with or accept others' work, but to take that work into account as part of the process of building the integrated body of reliable knowledge, perspective and method that constitutes the field of research in undergraduate mathematics education.

REVIEW PROCEDURES.

All papers, including invited submissions, will be evaluated by a minimum of three referees, one of whom will be a Volume editor. Papers will be judged on the basis of their originality, intellectual quality, readability by a diverse audience, and the extent to which they serve the pure and applied purposes identified earlier.

SUBMISSIONS.

Papers of any reasonable length will be considered, but the likelihood of acceptance will be smaller for very large manuscripts.

Five copies of each manuscript should be submitted. Manuscripts should be typed double-spaced, with bibliographies done in the style established by the American Mathematical Society for its CBMS series of volumes (an example style sheet is available from the editors on request).

Note that the *RCME* volumes are produced for electronic submission to the AMS. Accepted manuscripts should be prepared using AMS-TeX 2.1 (the macro packages are available through e-mail without charge from the AMS). Illustrations should also be prepared in a form suitable for electronic submission (for example, encapsulated postscript files).

CORRESPONDENCE.

Manuscripts and editorial correspondence should be sent to one of the three editors:

Ed Dubinsky
Department of Mathematics
Purdue University
West Lafayette, IN 47907
(bbf@j.cc.purdue.edu)

James Kaput
Department of Mathematics
University of Massachusetts, Dartmouth
North Dartmouth, MA 02747
(JKAPUT@umassd.edu)

Alan Schoenfeld
School of Education
University of California
Berkeley, CA 94720
(alans@violet.berkeley.edu)